Student Solutions M

for

Devore and Peck's
Statistics
The Exploration and Analysis of Data

Fifth Edition

and

Peck, Olsen, and Devore's
Introduction to Statistics and Data Analysis

Second Edition

Mary Mortlock
California Polytechnic State University

THOMSON

BROOKS/COLE

Australia • Canada • Mexico • Singapore • Spain • United Kingdom • United States

Printed in Canada
2   3   4   5   6   7   08   07   06   05   04

Printer: Webcom Limited

ISBN: 0-534-46522-6

For more information about our products, contact us at:
Thomson Learning Academic Resource Center
1-800-423-0563

For permission to use material from this text or product, submit a request online at
http://www.thomsonrights.com.
Any additional questions about permissions can be submitted by email to **thomsonrights@thomson.com.**

Thomson Brooks/Cole
10 Davis Drive
Belmont, CA 94002-3098
USA

Asia
Thomson Learning
5 Shenton Way #01-01
UIC Building
Singapore 068808

Australia/New Zealand
Thomson Learning
102 Dodds Street
Southbank, Victoria 3006
Australia

Canada
Nelson
1120 Birchmount Road
Toronto, Ontario M1K 5G4
Canada

Europe/Middle East/South Africa
Thomson Learning
High Holborn House
50/51 Bedford Row
London WC1R 4LR
United Kingdom

Latin America
Thomson Learning
Seneca, 53
Colonia Polanco
11560 Mexico D.F.
Mexico

Spain/Portugal
Paraninfo
Calle/Magallanes, 25
28015 Madrid, Spain

Table of Contents

Chapter 1

Exercise 1.1 – 1.7

1.1   Descriptive statistics is made up of those methods whose purpose is to organize and summarize a data set. Inferential statistics refers to those procedures or techniques whose purpose is to generalize or make an inference about the population based on the information in the sample.

1.3   The population of interest is the entire student body (the 15,000 students). The sample consists of the 200 students interviewed.

1.5   The population consists of all single-family homes in Northridge. The sample consists of the 100 homes selected for inspection.

1.7   The population consists of all 5000 bricks in the lot. The sample consists of the 100 bricks selected for inspection.

Exercise 1.8 – 1.19

1.9
   a   categorical
   b   categorical
   c   numerical (discrete)
   d   numerical (continuous)
   e   categorical (each zip code identifies a geographical region)
   f   numerical (continuous)

1.11   Possible values would include:
   a   General Motors, Toyota, Aston Martin, Ford, Jaguar, …..
   b   3.23, 2.92, 4.0, 2.8, …..
   c   2, 0, 1, 4, 3, ….
   d   49.2, 48.84, 50.3, 50.23, …..
   e   10, 15.5, 17, 3, 6.5, …….

1.13   **a.**

Grade	Frequency	Relative Frequency
A+	11	0.306
A	10	0.278
B	3	0.083
C	4	0.111
D	4	0.111
F	4	0.111
Total	36	1.000

Water quality ratings of California Beaches

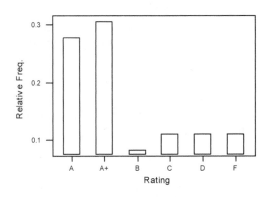

Assuming that an A+ means little risk of getting sick, most beaches in California seem quite safe. Two thirds of the beaches are rated a B or higher.

b.   No, a dotplot would not be appropriate. "Rating" is categorical data and a dotplot is used for small numerical data sets.

1.15

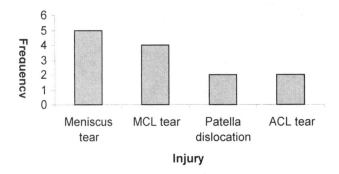

Knee injuries Amongst 13 Women Collegiate Rugby Players

The more common injury amongst the 13 female rugby players is the meniscus tear and a MCL tear. The incidence of both of these is at least double that of an ACL tear or patella dislocation.

1.17 The relative frequencies must sum to 1, so since .40+.22+.07 = .69, it must be that 31% of those surveyed replied that sleepiness on the job was not a problem.

Sleepy at Work?	Relative Frequency
Not at all	0.31
Few days each month	0.40
Few days each week	0.22
Daily Occurrence	0.07

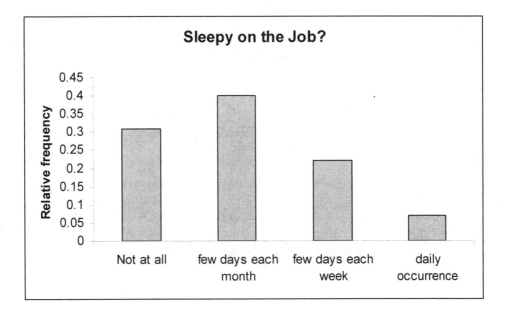

1.19

Most U.S. business schools have acceptance rates of between 16% and 38%. On school has a slightly lower rate than this (12%) and three schools have a much higher acceptance rate (between 43& and 49%) than the rest of the schools.

Exercise 1.20 – 1.24

1.21

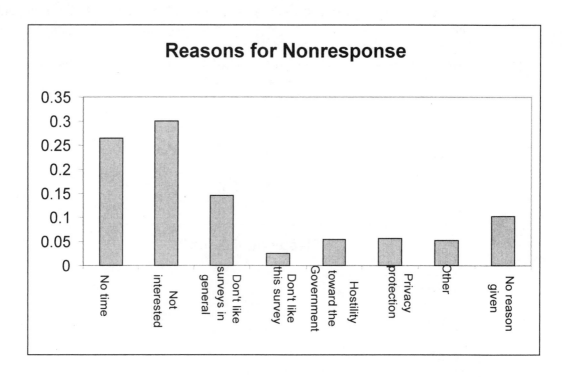

1.23   The display suggests that a representative value is around .93. The 20 observations are quite spread out around this value. There is a gap that separates the 3 smallest and the 3 largest cadence values from the rest of the data.

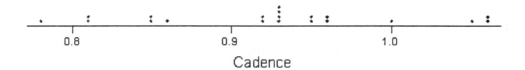

Chapter 2

Exercises 2.1 – 2.26

2.1   The first method is as follows. Write the graduates' names on slips of paper (one name per slip). Place the slips of paper in a container and thoroughly mix the slips. Then, select 20 slips (graduates), one at a time without replacement. The graduates whose names are on the slips constitute the random sample.

A second method would be to associate with each graduate a number between 1 and 140. Each graduate is assigned one number and no two graduates are assigned the same number. Then, use a random number generator on a computer to select 20 distinct random numbers between 1 and 140. Graduates whose numbers are selected are included in the sample.

2.3   The names appear on the petition in a list ordered from first to last. Using a computer random number generator, select 30 random numbers between 1 and 500. Signatures corresponding to the 30 selected numbers constitute the random sample.

2.5   Both procedures are unbiased procedures. It would be helpful to consider any known patterns in the variability among trees. Different rows may be exposed to different degrees of light, water, insects, nutrients, etc. which may affect the sugar content of the fruits. If this is true, then Researcher A's method may not produce a sample that is representative of the population. However, if the rows and trees are homogeneous, then the convenience and ease of implementation of Researcher A's method should be considered.

2.7   As the students are already in a numbered list, we can assign each of them their number on the list. As the largest number used is 4 digits, we will express all the numbers as 4 digits long. e.g. # 5 on the list will be assigned the number 0005. # 342 on the list will be assigned the number 0342. Using a random number table, we can go along a row taking each 4 digit number, discarding any repeats and any numbers greater than 4500 until a sample of 20 numbers have been obtained. The 20 students corresponding to these numbers on the list will become the sample.

2.9 Using a convenience sample may introduce selection bias. Suppose I wanted to know the average heights of students in all Stats classes, and used the students in my 10:00 class as my subjects. It may be that all the basketball players (that are generally tall) are in my 10:00 class because they have an early practice. Not only would my non-basketball 8:00 students be excluded from being part of the sample, but resulting average height would over-estimate the average height of my students. The sample must be representative of the population.

2.11 **a.** Using a random number generator, create 3 digit numbers. The 3 digits represent the page number, and the number of correctly defined words can be counted. Suppose we want to choose 10 pages in the book. Repeat this process until 10 page numbers are selected and count the number of "words" on each page.

      b. As the number of words on a page may be influenced by which topic being discussed – and therefore which chapter, it makes sense to stratify by chapter. Suppose we want to sample 3 pages in each chapter. For the first chapter and using a random number table, generate a 2 digit number. Use this to select a page within the chapter. Repeat this process twice more until 3 pages are chosen. Find these page numbers in the chapter and count the number of defined words on that page. Do this for each chapter until 3 pages from each chapter are chosen.

      c. Choose the 10th page and then every 15th page after that.

      d. Randomly choose one Chapter and count the words on each page in that chapter.

      e.&f. answers will vary

2.13 In theory, this is a good example of stratified random sample. Randomization is used in selecting the households in California. Another random technique is used to randomize the available subjects in each family. However, there are several possible sources of bias. It will eliminate households without a phone. It may be that the person who has recently had a birthday may be busy or too grumpy to talk to someone and some-one else (who may not be of voting age) responds to the questions. Someone may lie and give responses that are not true.

2.15 The best method to use would be to use a simple random sample where every different sample has an equal chance of being selected. For example, if each student in the school is numbered, you could use a random number generator to select a random sample of them. you could then find them, ask their opinion and then assume their views are representative of the whole schools.

2.17   Any survey conducted by mail is subject to selection bias; it eliminates any-one who doesn't have a permanent address, any-one who is on vacation or for any other reason doesn't receive mail. Once they receive the survey, many people consider surveys as junk mail and only respond to those that elicit strong feelings - resulting in non-response bias (only a few people reply).

2.19   Selection bias. The studies looked at women who had sisters, mothers and grandmothers who already had breast cancer.

      b   Time of day should be varied so as not to exclude those who work or those who work non-typical hours. Each selected random phone number may have to be dialed during several different hours of the day to maximize the chance of contacting the corresponding individual.

2.21   **a**   The population is all 16-24 year olds that live in New York state but not in New York City.

      b   Since the sample selection excluded New York City residents, we should not generalize to this group. Since fewer New York City young adults drive, the behavior of individuals who live in New York City could differ in important ways from that of individuals in the population sampled.

      c   Since the sample selection included only New York residents and only those aged 16-24, it would be unwise to generalize to other age groups or locations.

2.23   Firstly, the sample consists of only women and male responses may be different from the women's responses. Secondly, the participants are all volunteers and volunteer responses usually differ from those who choose not to participate. And thirdly, the participants are all from the same university which may not be representative of the entire nationwide college population.

2.25   This is an example of selection bias. The example states that 26% of Americans have Internet access. This means that a large portion (74%) of the population have no chance of being included in a public opinion poll. If those who are excluded from the sampling process differ from those who are included, with respect to the information that is being collected, then the sample will not be representative of the population.

Exercises 2.27 – 2.37

2.27   Pregnant women are interviewed and asked if they suffered severe morning sickness early in pregnancy. The baby's sex is noted. This scientific study is an observational study because we are interested in answering questions about characteristics of an existing population. The condition of morning sickness could not be controlled by the experimenter.

2.29   This is an observational study – the diabetes (the treatment) was not imposed by the experimenter and so the results were simple being observed. No cause-and-effect conclusion can be made on the basis on an observational study.

2.31   If we are considering all people 12 years of age or older, there are probably more single people that are young, and more widowed people that are old. It tends to be the young who are at higher risk of being victims of violent crimes (for instance, staying out late at night). Hence, age could be a potential confounding variable.

2.33   **a.**   If the definition of affluent Americans was having a household income of over $75,000 and that the sample was a simple random sample.

   b.   No. This sample is not representative of all Americans, since only affluent Americans were included.

2.35   The article does not report if SUV owners wear seat belts. It states that 98% of those injured or killed were not wearing seat belts. It may be that most SUV owners do not wear seat belts, which would explain the high rate of deaths by ejection.

2.37   It is possible that other confounding variables may be affecting the study's conclusion. It could be that men who eat more cruciferous vegetables are also making a conscious choice about eating healthier foods. A definitive causal connection cannot be made based on an observational study alone.

Exercises 2.38 – 2.48

2.39   The response variable is the amount of time it takes a car to exit the parking space.

   Extraneous factors that may affect the response variable include: location of the parking spaces, time of day, and normal driving behaviors of individuals. The factor of interest is whether or not another car was waiting for the parking space.

   One possible design for this study might be to select a frequently used parking space in a parking lot. Station a data collector at that parking space for given time period (say 2pm-5pm) every day for several days. The data collector will record the amount of time the driver uses to exit the parking space and whether another car was waiting for that space. The data collector will also note any other factors that might affect the amount of time the driver uses to exit such as buckling children into their car seats. Additionally, this same setup should be replicated at several different locations.

2.41   So many other factors could have contributed to the difference in the pregnancy rate. The only way that the difference between the two groups could have been attributed to the program was if the 2 groups were originally formed by dividing all the students completely randomly. This would minimize the effects of any other factors leaving the program the only <u>big</u> difference between the 2 groups.

2.43   An experiment that fairly compares the word processing programs while simultaneously accounting for possible differences on user's computer proficiency would have the same user testing each one of the four new word processing programs. The design would randomly select the order that each user tests the software. This would minimize the effects any extraneous factors (such as possible tiredness, boredom or frustration with the software) might have on speed with which the tasks are completed.

2.45   Yes, blocking on gender is useful for this study because 'Rate of Talk' is likely to be different for males and females.

2.47   Since all the conditions under which the experiment was performed are not given in this problem, it is possible that there are confounding factors in the experiment. Such factors might be the availability of cigarettes, the odor of cigarettes in the air, the presence of ashtrays, the availability of food, or magazines in the room that contain cigarette ads. Any of these factors could explain the craving for cigarettes. Assuming that the researchers were careful enough to control for these extraneous factors, the conclusion of the study would appear to be valid.

Exercises 2.49 – 2.58

2.49   A placebo treatment if often used to see if there is a psychological response in humans to a treatment. For instance, to test a new drug to reduce stress, by simply taking a sugar pill (the placebo) may reduce stress. A control group, made up of people in the same circumstances as those in the experimental and placebo groups would indicate if there was any other reason that stress might have been reduced; for instance a big holiday season or an event at work

2.51   It is necessary to divide into groups that are similar by a certain factor and eliminate any differences in responses. For examples, if you suspect the reaction to a drug may be different between the genders, it would be better to block by gender first before randomly dividing into the treatment groups and the place groups.

2.53   **a**   There are several extraneous variables, which could affect the results of the study. Two of these are subject variability and trainer variability. The researcher attempted to hold these variables constant by choosing men of about the same age, weight, body mass and physical strength and by using the same trainer for both groups. The researcher also included replication in the study.

Ten men received the creatine supplement and 9 received the fake treatment. Although the article does not say, we hope that the subjects were randomly divided between the 2 treatments.

b   It is possible that the men might train differently if they knew whether they were receiving creatine or the placebo. The men who received creatine might have a tendency to work harder at increasing fat-free mass. So it was necessary to conduct the study as a blinded study.

c   If the investigator only measured the gain in fat-free mass and was not involved in the experiment in any other way, then it would not be necessary to make this a double blind experiment. However, if the investigator had contact with the subjects or the trainer, then it would be a good idea for this to be a double blind experiment. It would be particularly important that the trainer was unaware of the treatments assigned to the subjects.

2.55   a.   No, the judges wanted to show that one of Pismo's restaurants made the best chowder.

b.   So that the evaluation is not swayed by personal interest.

2.57   a.   Randomly divide volunteers into 2 groups of 50, one groups gets PH80 and the other group gets a placebo nasal spray. They are assessed before and after the treatment and any improvement in PMS symptoms measured. It would be more accurate if neither the subjects or the assessors and recorders of the PMS symptoms knew which treatment group each subject had been assigned.

b.   A placebo treatment is needed to see if improvement is due to the PH80 or just a the act of spraying a liquid (with no medicinal qualities) up your nose that improves the symptoms of PMS.

c.   As irritability is so subjective, double-blinding, as described in **a**, would be advisable.

Exercises 2.59 – 2.64   Answers will vary

Exercises 2.65 – 2.72

2.65   **a**   By stratifying by province, information can be obtained about individual provinces as well as the whole country of Canada. Also, alcohol consumption may differ by province, just like we expect differences among states in the US.

b   Occupation is one socioeconomic factor that could be used for stratification. Alcohol consumption habits may be different based on a person's job. For example, a corporate businessman is likely to have more corporate sponsored social events involving alcohol consumption than a day care worker. Yearly income is another factor to use for stratification. Since alcoholic drinks are not free, those people with a high yearly income are likely to be able to afford the alcoholic drink of their choice.

2.67   The manager should number each room. For a simple random sample, he should write the number of each room on separate sheets of paper, put all the numbered papers into a hat and randomly select 15 papers. Those are the numbers for the 15 rooms he should survey. For a stratified sample, he should again write the number of each room on separate pieces of paper, and then divide the papers into piles entitled economy, business class and suites. Randomly choose 5 rooms from each grouping.

A stratified random sample would be more appropriate because the quality of housekeeping may vary from one room type to the next, but remain fairly similar within a given type. Possible extraneous variables include differences in cleaning style among housekeepers, room arrangements (some rooms may be harder to clean than others), and hotel occupancy rate (housekeepers may be rushed when the hotel is full).

2.69   Divide the 500m square plot into 4 equal size subplots, each measuring 250m x 250m, using two rows and two columns. Now divide each subplot again into 4 equal size smaller plots, each measuring 125m x 125m, using the same pattern. The result is, the 500m square plot is divided into 16 subplots with 4 rows of 4 subplots in each row. Now arrange the 4 types of grasslands so that each type appears in every row and column and in every 2x2 subplot. This is done to allow for repetition for each treatment (different grasslands). List all possible arrangements such that these conditions are held, and randomly select one to use in the experiment. Randomization is used in selecting the type of grassland arrangement for the plot as an effective way to even out the influences of extraneous variables. A few of the possible confounding variables in this experiment include exposure to sun or shade, proximity to water, slope of the land or possibly the number of worms in the soil. This study is an experiment since we are observing how a response variable (nesting) behaves when one or more factors (grasslands) are changed.

2.71   There are many possible designs any of which would be a reasonable design. Here is one example. Assume that the paper is published six days a week and assume that the lender is willing to advertise for six weeks. For the first week randomly select two days of the week on which advertisement one will be run. Then select randomly two days from the remaining four on which advertisement two will be run, and run advertisement three on the remaining two days. Repeat this randomization process for each week of the experiment. If the newspaper has two sections in which an advertisement can be placed, then randomly select three of the weeks and place the advertisement in section one, with the advertisement being run in section two during the remaining three weeks. The randomizations described should control the extraneous factors such as day of week, section of paper, and daily fluctuations of interest rates.

Chapter 3

Exercises 3.1 to 3.16

3.1

View of U.S. Economy

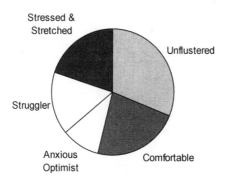

3.3   **a**

Same Salary for Same Work?

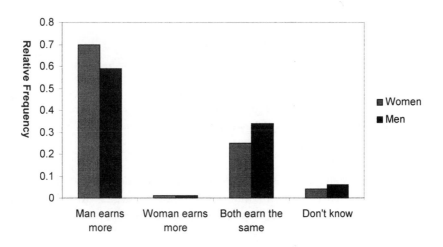

b

Women's views

Men's views

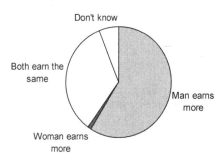

c   It's much easier to compare the differences using the comparative bar chart. The categories are next to each other and any difference is easy to see. To compare the size of the slice between two pie charts is more difficult.

d   The majority of both men and women think that men earn more money for the same work. Very few men or women think that a woman earns more. More women than women think that a man earns more or earns the same as woman.

3.5   **a**

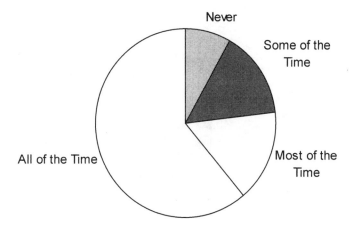

b

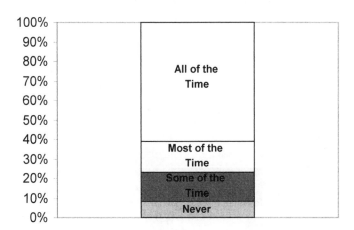

c   It is easier to see the difference in the proportions with a pie chart, however it is easier to estimate the percentage of each response with a segmented bar chart.

Who would you prefer to work for?

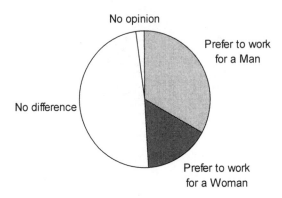

Given the choice, most people don't mind whether they work for a man or a woman. Of those that do have an opinion, most would prefer to have a male boss.

b

Would would you prefer to work for?

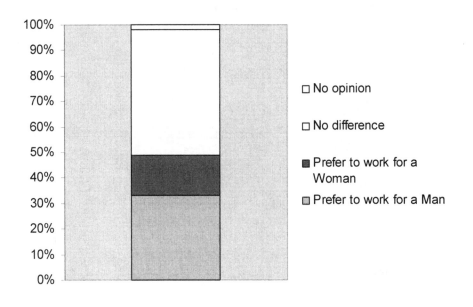

3.9

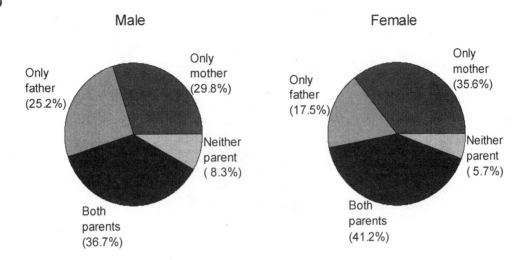

Male

Only father (25.2%)

Only mother (29.8%)

Neither parent (8.3%)

Both parents (36.7%)

Female

Only father (17.5%)

Only mother (35.6%)

Neither parent (5.7%)

Both parents (41.2%)

The pie charts show that a very large proportion of college students with medical career aspirations have at least one parent who is a physician. The charts also show that the female college students in this study have more mothers who are physicians than the male college students in the study have fathers who are physicians.

3.11   a

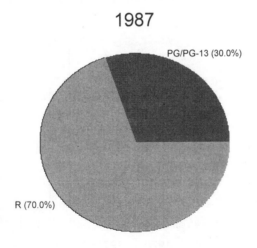

1987

PG/PG-13 (30.0%)

R (70.0%)

1992

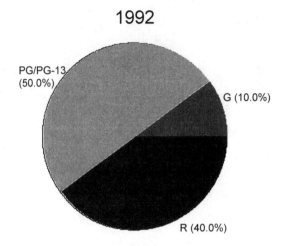

In 1987, 70% of the top ten movies that made the most money were rated R. The remaining 30% were rated G. By 1992, those percentages had changed. In that year, 40% of the top ten movies that made the most money were rated R, 50% were rated PG/PG-13 and 10% were rated G.

3.13

% of U.S. gross domestic product spent on Healthcare

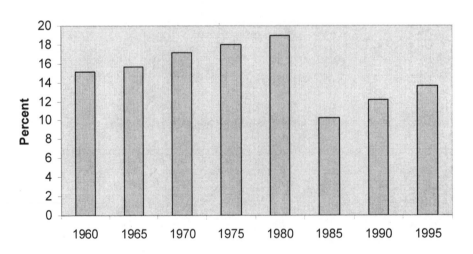

The bar graph shows a steady increase in the percent of US gross domestic product spent on health care over the 1960-1980 time period. There was then a sharp drop of about 8.5% in 1985 but increased again over the next 10 years. In 1995 the percent spent on health care was still less than that spent in 1960.

3.15   A bar chart would be better. There are too many different categories to use a pie chart. Even so, it might be worth considering collapsing some of the categories with smaller sales.

Sales of Major Retailing Categories

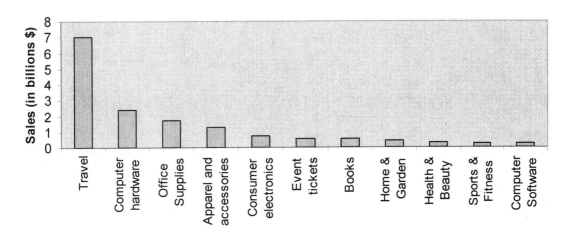

Exercises 3.17 to 3.24

3.17   **a**   2.2 liters/min.

   b   In the row with stem of 8. The leaf of 9 would be placed to the right of the other leaves.

   c   A large number of flow rates are between 6.0 and 8.0. Perhaps 6.9 or 7.0 could be selected as a typical flow rate.

   d   There appears to be quite a bit of variability in the flow rates. While there are a large number of flow rates in the 6.0 to 9.0 range, the flow rates appear to vary quite a bit in relation to one another.

   e   The distribution is not symmetric. Taking 7.0 as a typical value, the smaller flow rates are spread from 2.2 to 7.0, while the larger flow rates are spread from 7.0 to 18.9 (a larger spread).

   f   The value 18.9 appears to be somewhat removed from the rest of the data and hence is an outlier.

3.19

	Calorie Content (cal/100ml)of 26 Brands of Light Beers
1L	
1H	9
2L	2 3
2H	7 8 8 9 9 9
3L	0 0 1 1 1 2 2 3 3 4 5
3H	9
4L	0 1 2 3
4H	

stem: tens

leaf: ones

3.21 **a**

	% of fully credentialed teachers in CA counties
7	5
8	0 3 3 4 5 5 5 5 5 6 7 8 8 8 9 9
9	0 0 1 1 1 1 2 2 3 4 4 4 4 5 5 5 5 5 5 5 5 5 6 6 7 7 7 7 7 7 7 7 8 8 8 8 8 8 9 9
10	0

stem: tens

leaf: ones

Most counties in California have over 90% of their teachers fully credentialed. Los Angeles County is the lowest with only 75% and Alpine is the only County where 100% of their teachers are fully credentialed.

b

	% of fully credentialed teachers in CA counties
7L	
7H	5
8L	0 3 3 4
8H	5 5 5 5 5 6 7 8 8 8 8 9 9
9L	0 0 1 1 1 1 2 2 3 4 4 4 4
9H	5 5 5 5 5 5 5 5 6 6 7 7 7 7 7 7 7 7 7 8 8 8 8 8 8 9 9
10L	0

stem: tens

leaf: ones

We can now see that there are only three counties with less then 85% of their teachers fully credentialed.

3.23 **a**

	% increase in population 1990 to 2000	
0	0 1 3 4 4 4 5 5 5 5 6 7 8 8 8 8 9 9 9 9 9	
1	0 0 0 0 0 0 1 1 2 3 4 4 4 4 5 7 8	stem: tens
2	0 0 1 1 3 3 6 8	leaf: ones
3	0 1	
4	0	
5		
6	6	

b   48 of the states have an increase in population of 31% or less, and most of these are under 12%. There are two states that have a much larger %increase: Nevada (66%), and Arizona (40%)

c

% increase in population 1990 to 2000

WEST		EAST
9 9 8 8 8 0	0	1 3 4 4 4 5 5 5 5 6 7 8 9 9 9
4 4 3 0	1	0 0 0 0 0 1 1 2 4 4 5 7 8
8 3 1 0 0	2	1 3 6
1 0	3	
0	4	stem: tens
	5	leaf: ones
6	6	

The States that show a large % increase in population are in the West. There are 5 states in the West (out of 19) that has a % increase greater than the maximum increase in the East.

Exercises 3.25 to 3.40

3.25 **a**

Number of Impairments	Frequency	Relative Frequency
0	100	.4167
1	43	.1792
2	36	.1500
3	17	.0708
4	24	.1000
5	9	.0375
6	11	.0458
	n = 240	1.0000

b   .4167 + .1792 + .1500 = .7459

c   1 − .7459 = .2541

d   .1000 + .0375 + .0458 = .1833

e   The frequencies (and relative frequencies) tend to decrease as the number of impairments increase.

3.27   **a**

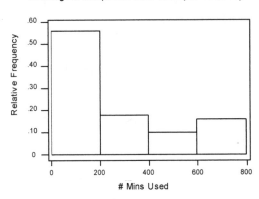

Average # Cell phone mins used per Month by Men

Most men average between 0 and 200 mins a month. Far fewer average between 400 and 800 mins a month.

b

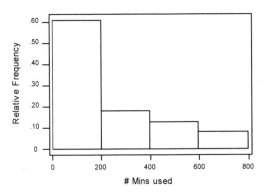

Average # Cell phone mins used per Month by Women

The distribution for men and women is similar in that most women average between 0 and 200 mins as do men. Fewer women average 600 to 800 minutes than men.

c   .18 + .56 = .74

d $.74 + \dfrac{.1}{2} = .79$

e $\dfrac{.13 * 3}{4} + .08 = .1775$

3.29   **a**   If a commute is longer than 45 minutes, then the time is often rounded to the nearest 15 or 30 minutes.

b

Commute Time	Frequency	Relative Frequency	Density
0 to < 5	5200	.0518	.0104
5 to < 10	18200	.1813	.0363
10 to < 15	19600	.1952	.0390
15 to < 20	15400	.1534	.0390
20 to < 25	13800	.1375	.0307
25 to < 30	5700	.0568	.0275
30 to < 35	10200	.1016	.0114
35 to < 40	2000	.0199	.0203
40 to < 45	2000	.0199	.0040
45 to < 60	4000	.0398	.0027
60 to < 90	2100	.0209	.0007
90 to < 120	2200	.0219	.0007

c

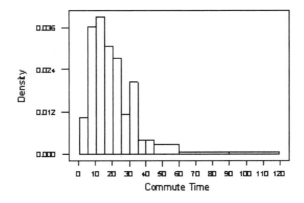

Most working adults commute between 10 minutes and an hour to work. However, there are a still a surprising number (look at the *area*) that commute between one and two hours to work.

d

Commute Time	Relative Frequency	Cumulative Relative Frequency
0 to < 5	.0518	.0518
5 to < 10	.1813	.2331
10 to < 15	.1952	.4283
15 to < 20	.1534	.5817
20 to < 25	.1375	.7192
25 to < 30	.0568	.7760
30 to < 35	.1016	.8776
35 to < 40	.0199	.8975
40 to < 45	.0199	.9174
45 to < 60	.0398	.9572
60 to < 90	.0209	.9781
90 to < 120	.0219	1.0

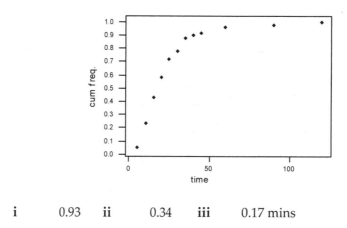

i   0.93   **ii**   0.34   **iii**   0.17 mins

3.31   **a**   We don't know the width of the last interval "100,000 or more" and the widths of the intervals are unequal.

b

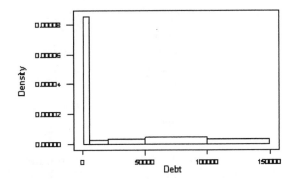

c   Only 42.7% of medical student who have completed their residencies have educational debts of under $5,000. 41.8% of them have educational debts of between $50,000 and $150,000. It seems that medical students finishing their residency have either a very large debt or a relatively small debt with only 15.5% of them having a debt of between $5,000 and $50,000.

3.33   a   If the exam is quite easy, then there would be a large number of high scores with a small number of low scores. The resulting histogram would be negatively skewed.

        b   If the exam were quite hard, then there would be a large number of low scores with a small number of high scores. The resulting histogram would be positively skewed.

        c   The students with the better math skills would score high, while those with poor math skills would score low. This would result in basically two groups and thus the resulting histogram would be bimodal.

3.35   a and b

Classes	Frequency	Rel. Freq.	Cum. Rel. Freq.
0 –< 6	2	.0225	.0225
6 –< 12	10	.1124	.1349
12 –< 18	21	.2360	.3709
18 –< 24	28	.3146	.6855
24 –< 30	22	.2472	.9327
30 –< 36	6	.0674	$1.0001 \approx 1.0$
	n = 50	1.0000	

Classes	Frequency	Rel. Freq.	Cum. Rel. Freq.
0 –< 6	2	.0225	.0225
6 –< 12	10	.1124	.1349
12 –< 18	21	.2360	.3709
18 –< 24	28	.3146	.6855
24 –< 30	22	.2472	.9327
30 –< 36	6	.0674	$1.0001 \approx 1.0$
	n = 50	1.0000	

c (Rel. Freq. for 12 –< 18) = (Cum. Rel. Freq. for < 18) – (Cum. Rel. Freq. for < 12)
$$= .3709 - .1349 = .2360.$$

d i The proportion that had pacemakers that did not malfunction within the first year equals 1 minus the proportion that had pacemakers that malfunctioned within the first year (12 months), which is $1 - .1349 = .8651$.

ii The proportion that required replacement between one and two years after implantation is equal to the proportion that had to be replaced within the first 2 years (24 months) minus the proportion that had to be replaced within the first year (12 months). This is $0.6855 - 0.1349 = 0.5506$.

e

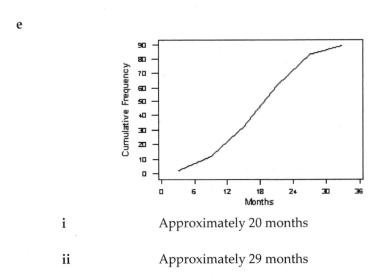

i Approximately 20 months

ii Approximately 29 months

3.37 Almost all the differences are positive indicating that the runners slow down. The graph is positively skewed. A typical difference value is about 150. About .02 of the runners ran the late distance more quickly that the early distance.

3.39

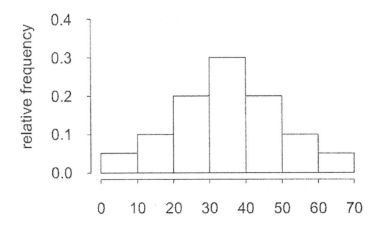

This histogram is symmetric.

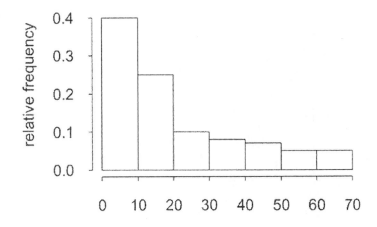

This histogram is positively skewed.

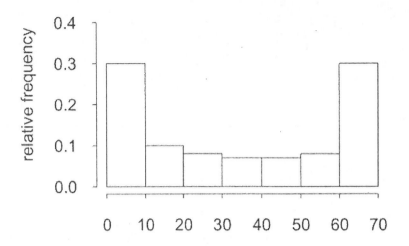

This is a bimodal histogram. While it is not perfectly symmetric it is close to being symmetric.

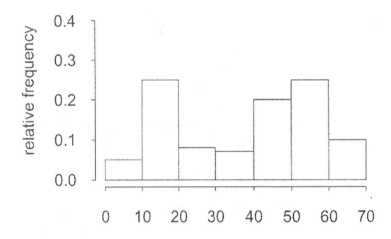

This is a bimodal histogram.

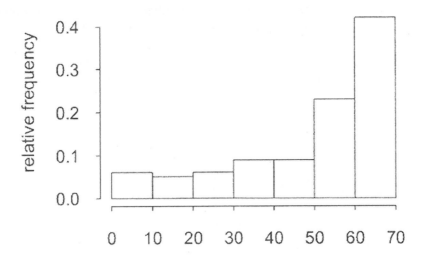

This is a negatively skewed histogram.

Exercises 3.41 to 3.53

3.41 **a** There are several values that have identical or nearly identical x-values yet different y-values. Therefore, the value of y is not determined solely by x , but also by various other factors. There appears to be a tendency for y to decrease as x increases.

b People with low body weight tend to be small people and it is possible their livers may be smaller than the liver of an average person. Conversely, people with high weight tend to be large people and their livers may be larger than the liver of an average person. Therefore, we would expect the graft weight ratio to be large for low weight people and small for high weight people.

3.43 **a.**

b. There is a positive relationship between the number of violations and the total fines. As the number of violations increases, so does the total fine.

3.45 **a**

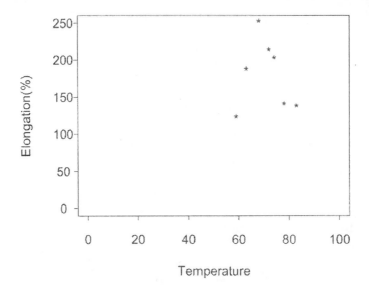

b

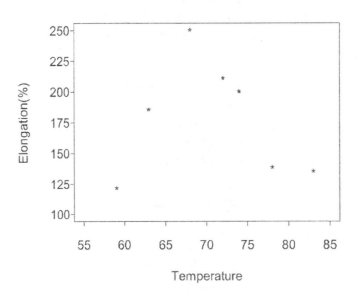

This plot is preferable to the plot in part **a**, because it reveals more clearly the possible curvilinear relationship between elongation and temperature.

c The graph suggests that elongation tends to increase as temperature increases until about seventy degrees and then elongation tends to decrease as temperature increases.

3.47 **a** For x = BOD mass, the median is $\dfrac{27+30}{2} = \dfrac{57}{2} = 28.5$.

The lower quartile is 11 and the upper quartile is 38.
The iqr is 38 – 11 = 27.
To check for outliers, we compute
1.5 × iqr = 1.5 × 27 = 40.5 and 3.0 × iqr = 3.0 × 27 = 81.
Therefore, 103 is a mild outlier and 142 is an extreme outlier.

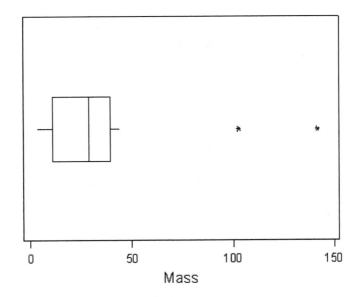

For y = BOD removal, the median is $\dfrac{11+16}{2} = \dfrac{27}{2} = 13.5$.

The lower quartile is 8 and the upper quartile is 30.
The iqr is 30 – 8 = 22.
To check for outliers, we compute
1.5 × iqr = 1.5 × 22 = 33.0 and 3.0 × iqr = 3.0 × 22 = 66.
Therefore, 75 and 90 are outliers.

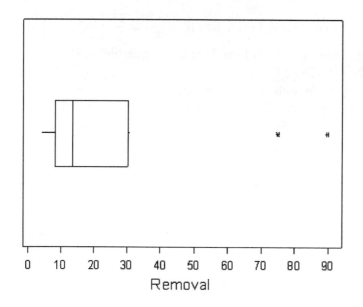

Removal

The medians are not centered for either of the boxplots suggesting skewness in the middle half of the data. There are 2 outliers for the BOD mass boxplot, one mild, 103 and one extreme, 142. There are also 2 outliers for the BOD removal boxplot, 75 and 90. Both of these outliers are mild.

b

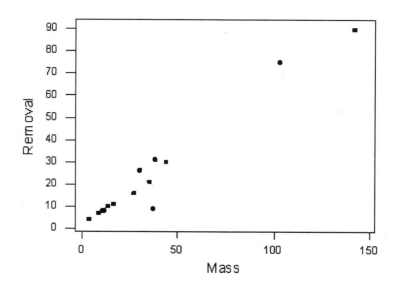

There is a tendency for y to increase as x increases. That is, the larger values for BOD mass will be associated with larger values for BOD removal.

3.49 **a**   There is not a deterministic relationship between x and y. This can be determined by the fact that there are two data points, (100, 222) and (100, 241), which have the same x-value but different y-values.

b

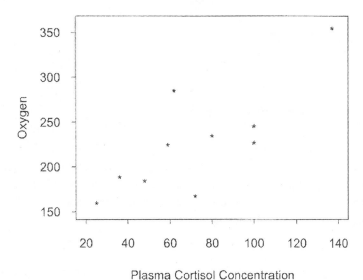

c   There appears to be a tendency for oxygen consumption rate to increase as plasma cortisol concentration increases.

3.51 **a**

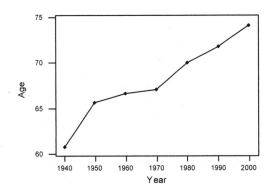

b

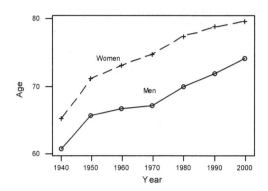

c   There has been a steady increase in life expectancy for both men and women from 1940 to2000. For both sexes, the life expectancy in 2000 is about 14 years higher than it was is in 1940, but the life expectancy for women is consistently about 6 years higher than that of men.

3.53   In both 2001 and 2002 the box office sales dropped in Weeks 2, 6 and in the last two weeks of the summer. The seasonal peaks occurred during Weeks 4, 9 and 13.

Exercises 3.54 to 3.66

3.55   **a**

Day of Week	Frequency
Sunday	109
Monday	73
Tuesday	97
Wednesday	95
Thursday	83
Friday	107
Saturday	100
	n = 664

b $\dfrac{(107+100+109)}{664}=\dfrac{316}{664}=.4759$, which converts to 47.59%.

c   If a murder were no more likely to be committed on some days than on other days, the proportion of murders on a specific day would be 1/7 = .1429. So, for three days the proportion would be 3(.1429) = .4287. Since the proportion for the weekend is .4759, there is some evidence to suggest that a murder is more likely to be committed on a weekend day than on a non-weekend day.

3.57   a

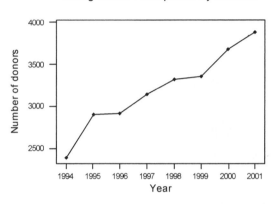

The number of transplants from a living relative has been increasing steadily from 1994 to 2001.

b

Living-donor Kidney Transplants

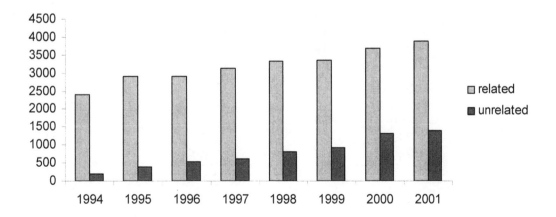

There are many more related donors than unrelated donors used for kidney transplants. However the use of non-related donors is rising at a faster rate than that of non-related donors.

3.59　**a**　The two histograms do give different impressions about the distribution of values. For the first histogram, it appears that more frozen meals have sodium content around 800mg. However, the second histogram suggests that sodium content is fairly uniform from 300mg to 900mg and then drops off above 900mg.

b　Using the first histogram, the proportion of observations that are less than 800 is approximately

$$\frac{6+7+\dfrac{10}{2}}{(6+7+10+4)} = \frac{18}{27} = .6667.$$

Using the second histogram, the proportion of observations that are less than 800 is approximately

$$\frac{6+5+5+\left(\dfrac{800-750}{150}\right)(6)}{(6+5+5+6+4+1)} = \frac{16+2}{27} = \frac{18}{27} = .6667.$$

The actual proportion is $\dfrac{18}{27} = \dfrac{2}{3} = .6667.$

3.61　**a**

Class	Frequency	Rel. Freq.
0 –< .1	0	.0000
.1 –< .2	0	.0000
.2 –< .3	0	.0000
.3 –< .4	6	.1304
.4 –< .5	13	.2826
.5 –< .6	14	.3043
.6 –< .7	5	.1087
.7 –< .8	3	.0652
.8 –< .9	3	.0652
.9 –< 1.0	0	.0000
1.0 –< 1.1	1	.0217
1.1 –< 1.2	1	.0217
	n = 46	.9998 ≈ 1.000

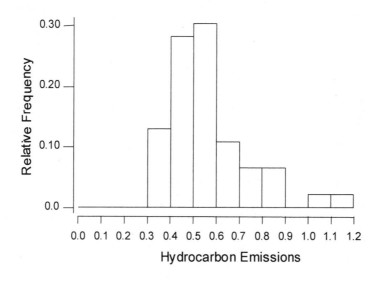

The HC values of 1.02 and 1.10 appear to be outliers.

b

Class	Frequency	Rel. Freq.
0 –< 2	1	.0217
2 –< 4	7	.1522
4 –< 6	15	.3261
6 –< 8	8	.1739
8 –< 10	3	.0652
10 –< 12	2	.0435
12 –< 14	2	.0435
14 –< 16	5	.1087
16 –< 18	0	.0000
18 –< 20	1	.0217
20 –< 22	0	.0000
22 –< 24	2	.0435
	n = 46	1.0000

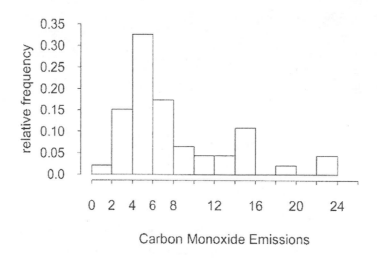

Carbon Monoxide Emissions

c   Both of the histograms are positively skewed.

3.63   **a**

Class Intervals	Frequency	Rel. Freq.
0 –< 0.5	5	.1064
.5 –< 1.0	9	.1915
1.0 –< 1.5	10	.2128
1.5 –< 2.0	9	.1915
2.0 –< 2.5	3	.0638
2.5 –< 3.0	5	.1064
3.0 –< 3.5	1	.0213
3.5 –< 4.0	3	.0638
4.0 –< 4.5	1	.0213
4.5 –< 5.0	1	.0213
	n = 47	1.0001

b

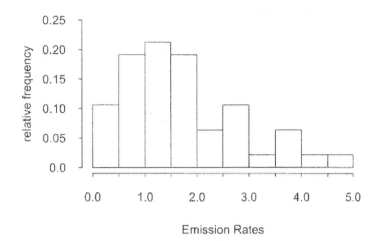

Emission Rates

The histogram is positively skewed.

c

Class Intervals	Cumulative Rel. Freq.
0 –< 0.5	.1064
.5 –< 1.0	.2979
1.0 –< 1.5	.5107
1.5 –< 2.0	.7022
2.0 –< 2.5	.7660
2.5 –< 3.0	.8724
3.0 –< 3.5	.8937
3.5 –< 4.0	.9575
4.0 –< 4.5	.9788
4.5 –< 5.0	1.0001

d **i** About .2979 or 29.79% of the states had SO_2 emission below 1.0.

   ii About .7022 − .2979 = .4043 or 40.43% of the states had SO_2 emission between 1.0 and 2.0.

   iii About 1 − .7022 = .2978 or 29.78% of the states has SO_2 emission which exceeded 2.0.

3.65   **a**

Babies	Frequency	Relative Frequency
1	1	.0059
2	2	.0118
3	11	.0647
4	21	.1235
5	35	.2059
6	38	.2235
7	33	.1941
8	18	.1059
9	8	.0471
10	2	.0118
11	1	.0059
	170	1.0001

b   The proportion of the litters with more than 6 babies = $\dfrac{33+18+8+2+1}{170}$ = .3647

The proportion with between 3 and 8 babies = $\dfrac{11+21+35+38+33+18}{170} = \dfrac{156}{170} =$

.9176

c   It is much easier to answer questions similar to those posed in (b) using the relative frequency table. The table gives the number of occurrences for each observation. This is the information needed to answer the questions. Using the raw data would require tallying for each question.

Chapter 4

Exercises 4.1 to 4.15

4.1   mean $= \dfrac{9707}{10} = 970.7$ median $= \dfrac{707+769}{2} = \dfrac{1476}{2} = 738$. There are a couple of outliers at the high end of the data set and so the mean is influenced by these more than the median. For this reason, the median is more representative of the sample.

4.3   The sample median determines this salary. Its value is $\dfrac{4443+4129}{2} = \dfrac{8572}{2} = 4286$.

The mean salary paid in the six counties is

$$\frac{5354+5166+4443+4129+2500+2220}{6} = \frac{23812}{6} = 3968.67.$$

Since the mean salary is less than the median salary, the mean salary is less favorable to the supervisors.

4.5   **a**   stem : tens   leaf : ones

stem	leaf
32	55
33	49
34	
35	6699
36	34469
37	03345
38	9
39	2347
40	23
41	
42	4

The stem and leaf display suggests that the mean and median will be fairly close to each other. Most values are between 356 and 375 and there are approximately equal amounts of values larger or smaller than these central values.

    b $\overline{x} = \dfrac{9638}{26} = 370.69$

          median $= \dfrac{369+370}{2} = 369.5$

c　　The largest value, 424, could be increased by any arbitrary amount without affecting the sample median. The median is insensitive to outliers. However, if the largest value was decreased below the sample median, 369.5, then the value of the median would change.

4.7　**10% trimmed mean is eliminating** $\dfrac{10}{100}(40) = 4$ points from each end: $\dfrac{430}{32} = 13.4375$

　　20% trimmed mean is eliminating $\dfrac{20}{100}(40) = 8$ points from each end: $\dfrac{302}{24} = 12.5833$

　　$\bar{x} = \dfrac{924}{40} = 23.1$　sample median = 13

4.9　The fact that only half the homes in this county cost less than \$278,380 is the definition of a median.

The median is not the midpoint of the range. It is the price that 50% of the homes cost less and 50% of the houses cost more than this price. If there are no houses costing below \$300,000 and unless at least half the homes in this county cost exactly \$300,000, Walker is correct; the median will be above \$300,000

4.11　The four observations ordered are: 27.3, 30.5, 33.5, 36.7.

The median = 32 and the mean = 128/4 = 32. If we add the value 32 to the data set, the new ordered data set ordered is 27.3, 30.5, 32, 33.5, 36.7. For this data set, the sample median is 32 and the mean is $\bar{x} = \dfrac{160}{5} = 32$.

4.13　The median and trimmed mean (trimming percentage of at least 20) can be calculated.

　　sample median = $\dfrac{57+79}{2} = \dfrac{136}{2} = 68$

　　20% trimmed mean = $\dfrac{35+48+57+79+86+92}{6} = \dfrac{397}{6} = 66.17$

4.15　The data arranged in ascending order is 4,8,11,12,33. The sample median is 11%. The large outlier, 33%, has pulled the mean out towards that outlying value. The median, on the other hand, is not affected by the outlier and hence gives a better indication of a typical return.

Exercises 4.16 to 4.29

4.17　**a**　　Set 1: 2, 3, 7, 11, 12 ; $\bar{x} = 7$ and $s = 4.528$

　　　　　　Set 2: 5, 6, 7, 8, 9 ; $\bar{x} = 7$ and $s = 1.581$

b   Set 1: 2, 3, 4, 5, 6 ; $\bar{x} = 4$ and $s = 1.581$

   Set 2: 4, 5, 6, 7, 8 ; $\bar{x} = 6$ and $s = 1.581$

4.19   mean $= \dfrac{20179}{27} = 747.37$ standard deviation $= 606.894$ (both in thousands of $)

   mean + 2 standard deviations $= 747.37 + 2(606.894) = 1961.158$ (or $1,961,158)

4.21   **a**   Moderately priced midsize cars: standard deviation: $406.98   iqr: $370

   b   Inexpensive midsize cars: standard deviation: $186.236 iqr: $155

   c   There is more variability in the repair cost for moderately priced midsize cars. Both the standard deviation and interquartile range values are higher than the corresponding values for the inexpensive midsize cars.

   d   mean repair cost for moderately priced midsize cars: $348.9
   mean repair cost for inexpensive midsize cars: $298.36

   e   On average, the repair cost is higher for a moderately priced midsize car compared to an inexpensive midsize car. The variability (or spread) of the individual costs is greater for the moderately priced cars.

4.23   **a**   mean: 2965.2   standard deviation: 542.6   iqr: 602

   b   As the iqr for the sodium content for chocolate pudding is 602 and the iqr for sodium content in catsup is 1300, there is less variability in sodium content in chocolate pudding.

4.25

x	x - x̄	(x - x̄)²
244	51.4286	2644.9009
191	-1.5714	2.4693
160	-32.5714	1060.896
187	-5.5714	31.0405
180	-12.5714	158.0401
176	-16.5714	274.6113
174	-18.5714	344.8969
205	12.4286	154.4701
211	18.4286	339.6133
183	-9.5714	91.6117
211	18.4286	339.6133
180	-12.5714	158.0401
194	1.4286	2.0409
200	7.4286	55.1841
2696	.0004	5657.4285

$\bar{x} = 192.5714$

$$s^2 = \frac{5657.4285}{13} = 435.1868$$

$s = 20.8611$

s^2 could be interpreted as the mean squared deviation from the average leg power at a high workload. This is 435.1868. The standard deviation, s, could be interpreted as the typical amount by which leg power deviates from the average leg power. This is 20.8611.

4.27 Subtracting 10 from each data point yields:

x	(x - x̄)	(x - x̄)²
52	13.636	185.9405
13	−25.364	643.3325
17	−21.364	456.4205
46	7.636	58.3085
42	3.636	13.2205
24	−14.364	206.3245
32	−6.364	40.5005
30	−8.364	69.9565
58	19.636	385.5725
35	−3.364	11.3165
73	34.636	1199.652

$$\bar{x} = \frac{422}{11} = 38.364$$

These deviations from the mean are identical to the deviations from the mean for the original data set. This would result in a variance for the new data set that is equal to the variance of the original data set. In general, adding the same number to each observation has no effect on the variance or standard deviation.

4.29   a   For sample 1, $\bar{x} = 7.81$ and $s = .39847$
                       For sample 2, $\bar{x} = 49.68$ and $s = 1.73897$

         b   For sample 1, $CV = (100)(.39847)/7.81 = 5.10$
                       For sample 2, $CV = (100)(1.73897)/49.68 = 3.50$

                     Even though the first sample has a smaller standard deviation than the second, the variation relative to the mean is greater.

Exercises 4.30 to 4.37

4.31

Number of times Web site Accessed

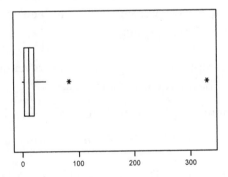

As with the dotplot, it is clear that there are 2 outliers at 84 and 331. Although individual values cannot be shown by a boxplot, it is clear that 75% of the class accessed the website 20 times or less during the first month.

4.33

Nitrous Oxide Emissions (thousands of tons)

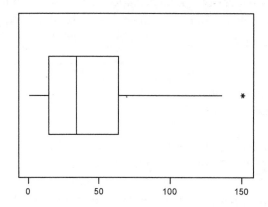

Half the states in the continental U.S. states have nitrous oxide emissions of less than 34000 tons. There is one state that has a much larger quantity of emissions than any of the other states The distribution looks negatively skewed meaning that most states have emission values at the lower end of the range.

4.35   The median score for Creamy is about 45 and the median score for Crunchy is about 51. The ranges appear to be equal. The data for Creamy is slightly skewed in the positive direction while the data for Crunchy is fairly symmetric.

4.37   **a**   For the excited delirium sample, the median is 0.4. The lower quartile is 0.1 and the upper quartile is 2.8. The interquartile range is 2.7.

For the No Excited Delirium sample, the median is $\frac{1.5+1.7}{2} = 1.6$. The lower quartile is 0.3 and the upper quartile is 7.9. The interquartile range is 7.6.

b   To check for outliers for the Excited Delirium sample, we must compute
1.5 x 2.7 = 4.05 and 3 x 2.7 = 8.1
The lower quartile – 4.05 = 0.1 – 4.05 = -3.05
The upper quartile + 4.05 = 2.8 + 4.05 = 6.85
The lower quartile – 8.1 = 0.1 – 8.1 = -8.0
The upper quartile + 8.1 = 2.8 + 8.1 = 10.9

So 8.9 and 9.2 are mild outliers and 11.7 and 21.0 are extreme outliers for the Excited Delirium sample.

To check for outliers for the No Excited Delirium sample, we must compute
1.5 x 7.6 = 11.4 and 3 x 7.6 = 22.8
The lower quartile –11.4 = 0.3-11.4 = -11.1
The upper quartile + 11.4 = 7.9 + 11.4 = 19.3
There are no outliers for this sample.

c

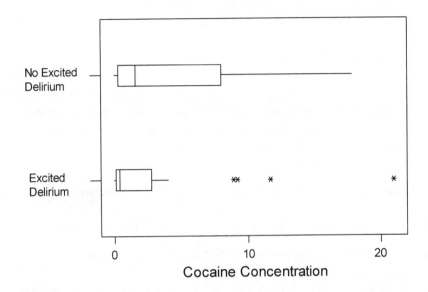

The median marker for each boxplot is located at the lower end of the boxplot. The boxplot for the No Excited Delirium is much wider than for Excited Delirium indicating much more variability in the middle half of the data. Both boxplots have short whiskers on the lower end and long whiskers on the higher end. The Excited Delirium sample has 2 mild outliers and 2 extreme outliers. There are no outliers for the No Excited Delirium sample.

Exercises 4.38 to 4.52

4.39   **a**   The value 57 is one standard deviation above the mean. The value 27 is one standard deviation below the mean. By the empirical rule, roughly 68% of the vehicle speeds were between 27 and 57.

   b   From part **a** it is determined that 100% – 68% = 32% were either less than 27 or greater than 57. Because the normal curve is symmetric, this allows us to conclude that half of the 32% (which is 16%) falls above 57. Therefore, an estimate of the percentage of fatal automobile accidents that occurred at speeds over 57 mph is 16%.

4.41   **a**   The value 59.60 is two standard deviations above the mean and the value 14.24 is two standard deviations below the mean. By Chebyshev's rule the percentage of observations between 14.24 and 59.60 is at least 75%.

b   The required interval extends from 3 standard deviations below the mean to 3 standard deviations above the mean. From $36.92 - 3(11.34) \doteq 2.90$ to $36.92 + 3(11.34) = 70.94$.

c $\bar{x} - 2s = 24.76 - 2(17.20) = -9.64$

In order for the histogram for NO_2 concentration to resemble a normal curve, a rather large percentage of the readings would have to be less than 0 (since $\bar{x} - 25 = -9.64$). Clearly, a reading cannot be negative, hence the histogram cannot have a shape that resembles a normal curve.

4.43   For the first test the student's z-score is $\dfrac{(625 - 475)}{100} = 1.5$ and for the second test it is $\dfrac{(45 - 30)}{8} = 1.875$. Since the student's z-score is larger for the second test than for the first test, the student's performance was better on the second exam.

4.45   Since the histogram is well approximated by a normal curve, the empirical rule will be used to obtain answers for part **a** – **c**.

a   Because 2500 is 1 standard deviation below the mean and 3500 is 1 standard deviation above the mean, about 68% of the sample observations are between 2500 and 3500.

b   Since both 2000 and 4000 are 2 standard deviations from the mean, approximately 95% of the observations are between 2000 and 4000. Therefore about 5% are outside the interval from 2000 to 4000.

c   Since 95% of the observations are between 2000 and 4000 and about 68% are between 2500 and 3500, there is about $95 - 68 = 27\%$ between 2000 and 2500 or 3500 and 4000. Half of those, $27/2 = 13.5\%$, would be in the region from 2000 to 2500.

d   When applied to a normal curve, Chebyshev's rule is quite conservative. That is, the percentages in various regions of the normal curve are quite a bit larger than the values given by Chebyshev's rule.

4.47   The recorded weight will be within 1/4 ounces of the true weight if the recorded weight is between 49.75 and 50.25 ounces. Now, $\dfrac{(50.25 - 49.5)}{.1} = 7.5$ and $\dfrac{(49.75 - 49.5)}{.1} = 2.5$.
Also, at least $1 - 1/(2.5)^2 = 84\%$ of the time the recorded weight will be between 49.25 and 49.75. This means that the recorded weight will exceed 49.75 no more than 16% of the time. This implies that the recorded weight will be between 49.75 and 50.25 no more than 16% of the time. That is, the proportion of the time that the scale showed a weight that was within 1/4 ounce of the true weight of 50 ounces is no more than 0.16.

4.49   Because the number of answers changed from right to wrong cannot be negative and because the mean is 1.4 and the value of the standard deviation is 1.5, which is larger than the mean, this implies that the distribution is skewed positively and is not a normal curve. Since
$(6 - 1.4)/1.5 = 3.07$, by Chebyshev's rule, at most $1/(3.07)^2 = 10.6\%$ of those taking the test changed at least 6 from correct to incorrect.

4.51   **a**

Class	Frequency	Rel Freq.	Cum. Rel. Freq.
5 –< 10	13	.26	.26
10 –< 15	19	.38	.64
15 –< 20	12	.24	.88
20 –< 25	5	.10	.98
25 –< 30	1	.02	1.00
	n = 50	1.00	

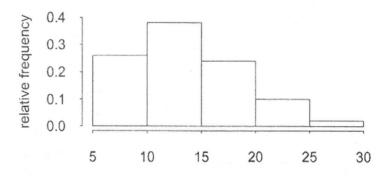

1989 per capita expenditures on libraries

b   **i**   The 50th percentile is between 10 and 15 since the cumulative relative frequency at 10 is .26 and at 15 it is .64. The 50th percentile is approximately
$10 + 5(50 - 26)/38 = 10 + 3.158 = 13.158$.

49

ii   The 70th percentile is between 15 and 20 since the cumulative relative frequency at 15 is .64 and at 20 it is .88. The 70th percentile is approximately
15 + 5(70 – 64)/24 = 15 + 1.25 = 16.25.

iii   The 10th percentile is between 5 and 10 and is approximately
5 + 5(10 – 0)/26 = 5 + 1.923 = 6.923.

iv   The 90th percentile is between 20 and 25 and is approximately
20 + 5(90 – 88)/10 = 20 + 1 = 21.

v   The 40th percentile is between 10 and 15 and is approximately
10 + 5(40 – 26)/38 = 10 + 1.842 = 11.842.

Exercises 4.53 to 4.68

4.53   **a.**   median: 20.88   1st Quartile: 18.09   3rd Quartile: 22.20

        b.   iqr = 22.2 – 18 09 = 4.11
1.5(iqr)= 1.5(4.11) = 6.165 Anything below 18.09 – 6.165 = 11.925 or above
22.2 + 6.165 = 28 365 would be an outlier. There are 2 outliers, 35. 78 and 36.73.

        c.

Oxygen capacity (mL/kg/min) for Male middle-aged Runners

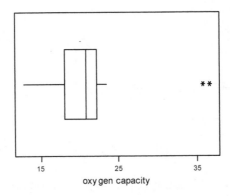

Most of the subjects have an oxygen capacity of between 20 and 25 mL/kg/min. There are 2 subjects with much higher oxygen capacities; of over 35 mL/kg/min.

4.55   **a** $\bar{x} = \dfrac{2696}{14} = 192.57$

The data arranged in ascending order is: 160, 174, 176, 180, 180, 183, 187, 191, 194, 200, 205, 211, 211, 244.

The median is the average of the two middle values since n = 14, an even number. The median value is $\dfrac{(187+191)}{2} = 189$.

      b   The median value would remain unchanged, but the mean value would decrease. The new mean value is $\dfrac{2656}{14} = 189.71$.

      c   The trimmed mean =

$\dfrac{174+176+180+180+183+187+191+194+200+205+211+211}{12} = \dfrac{2292}{12} = 191$. The

trimming percentage is $\dfrac{1}{14}$ x 100 = 7.14%

      d   If the largest observation is 204, then the new trimmed mean =

$\dfrac{174+176+180+180+183+187+191+194+200}{9} = \dfrac{1665}{9} = 185$

If the largest observation is 284, the trimmed mean is the same as in **c.**

4.57   **a**   The median is $\dfrac{7.93+7.89}{2} = \dfrac{15.82}{2} = 7.91$

The upper quartile is 8.01.
The lower quartile is 7.82
The interquartile range is 8.01 − 7.82 = 0.19.

      b   To check for outliers, we calculate

$1.5 \times$ iqr = $1.5 \times$ 0.19 = 0.285
$3.0 \times$ iqr = $3.0 \times$ 0.19 = 0.57
The upper quartile + 1.5iqr = 8.01 + 0.285 = 8.295
The lower quartile − 1.5iqr = 7.82 − 0.285 = 7.535
The upper quartile + 3.0iqr = 8.01 + 0.57 = 8.58
The upper quartile − 3.0iqr = 7.82 − 0.57 = 7.25

There are no outliers in this data set.

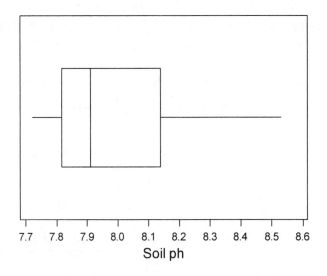

Soil ph

4.59　a

	Cancer		No Cancer
	9 6 8 3 7 9 5	0	9 5 7 6 8 3 9 7 6 7 8 9 9 3
8 6 0 7 1 8 1 5 0 6 6 8 1 5 2 3 3 1 5 0	1	1 2 2 7 1 7 1 3 1 1 4
	1 2 3 0 2 7 3 1	2	9 9 4 9 4 1 9 1
	8 3 4 9	3	8 3 9
	5	4	
	7	5	5 5
		6	
		7	
HI: 210		8	5

Both data sets are positively skewed, but the displays differ in the first two stems.

b　For the cancer group: $\bar{x} = \dfrac{958}{42} = 22.81$ and the median $= \dfrac{(16+16)}{2} = 16.$

For the no-cancer group: $\bar{x} = \dfrac{747}{39} = 19.15$ and the median $= 12.$

The values of both the mean and median suggest that the center of the cancer sample is to the right (larger in value) of the center of the no-cancer group.

c　For the cancer group:

$$s^2 = \frac{S_{xx}}{n-1} = \frac{41084.476}{41} = 1002.0604$$

$$s = \sqrt{1002.0604} = 31.6553$$

52

For the no-cancer group:

$$s^2 = \frac{S_{xx}}{n-1} = \frac{10969.077}{38} = 288.6599$$

$$s = \sqrt{288.6599} = 16.99$$

The values of s^2 and s suggest that there is more variability in the cancer sample values than in the no-cancer sample values.

d   For the cancer group, the iqr = 22 − 11 = 11.
For the no-cancer group, the iqr = 24 − 8 = 16.
The previous conclusion about variability is not confirmed by the interquartile ranges. A possible explanation is the extreme outlier in the cancer sample whose value is 210. Eliminating the largest value in each group and recalculating the variances yields the following results.

For the cancer group: $n = 41$, $\sum x = 748$, $\sum x^2 = 18836$

$$s^2 = \frac{18836 - \dfrac{(748)^2}{41}}{40} = \frac{18836 - 13646.439}{40} = \frac{5189.561}{40} = 129.739$$

$$s = \sqrt{129.739} = 11.39$$

For the no-cancer group: $n = 38$, $\sum x = 662$, $\sum x^2 = 18052$

$$s^2 = \frac{18052 - \dfrac{(662)^2}{38}}{37} = \frac{18052 - 11532.7368}{37} = \frac{6519.2632}{37} = 176.1963$$

$$s = \sqrt{176.1963} = 13.27$$

The values of s for the two groups with the outliers removed are closer in value and the no-cancer group is higher.

e

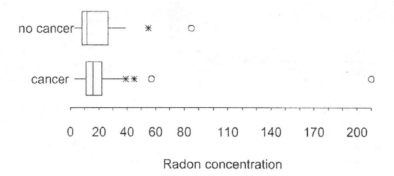

4.61 It implies that the median salary is less than $1 million and that the histogram would have a shape which is positively skewed.

4.63 **a**

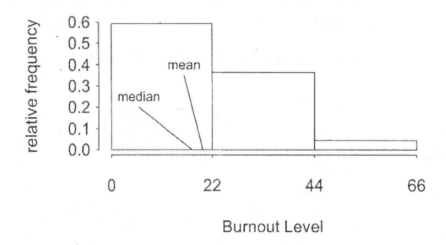

b The two measures of center are not identical because the histogram is not symmetric.

c Using Chebyshev's rule, we are guaranteed to find at least 75% of the scores in the interval extending from

$$\bar{x} - 2s = 19.93 - 2(12.89) = -5.85 \, to \, \bar{x} + 2s = 19.93 + 2(12.89) = 45.71.$$

Since a burnout level score cannot be negative, we could change the left-hand endpoint of –5.85 to 0. Therefore, we are guaranteed to find at least 75% of the scores in the interval extending from 0 to 45.71. From the histogram, about (554+342)/937 = .956, or roughly 96% fall in this interval.

4.65

	Smoking	Nonsmoking
$\overline{x}$	693.364	961.091
median	693.000	947.000
lower quartile	598.000	903.000
upper quartile	767.000	981.000
s^2	10787.700	12952.700
s	103.864	113.810

Boxplots for the two groups are below.

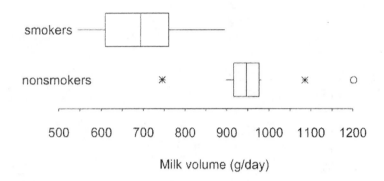

The data strongly suggests that milk volume for nonsmoking mothers is greater than that for smoking mothers.

4.67

x	$(x-\bar{x})$	$(x-\bar{x})^2$
18	−4.15	17.22
18	−4.15	17.22
25	2.85	8.12
19	−3.15	9.92
23	0.85	0.72
20	−2.15	4.62
69	46.85	2194.92
18	−4.15	17.22
21	−1.15	1.32
18	−4.15	17.22
18	−4.15	17.22
20	−2.15	4.62
18	−4.15	17.22
18	−4.15	17.22
20	−2.15	4.62
18	−4.15	17.22
19	−3.15	9.92
28	5.85	34.22
17	−5.15	26.52
18	−4.15	17.22
443	0.00	2454.48

a $\bar{x} = \dfrac{443}{20} = 22.15$

$s^2 = 2454.48/19 = 129.183$ and $s = \sqrt{129.183} = 11.366$

b   The 10% trimmed mean is calculated by eliminating the two largest values (69 and 28) and the two smallest values (17 and 18). The trimmed mean equals $311/16 = 19.4375$. It is a better measure of location for this data set since it eliminates a very large value (69) from the calculation. It is almost 3 units smaller than the average.

c   The upper quartile is $(21 + 20)/2 = 20.5$, the lower quartile is $(18 + 18)/2 = 18$, and the iqr $= 20.5 − 18 = 2.5$.

d   upper quartile + 1.5(iqr) = $20.5 + 1.5(2.5) = 24.25$
upper quartile + 3.0(iqr) = $20.5 + 3(2.5) = 28.00$
The values 25 and 28 are mild outliers and 69 is an extreme outlier.

e

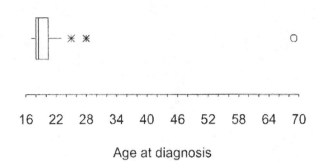

Age at diagnosis

Chapter 5

Exercises 5.1 – 5.16

5.1 **a** A positive correlation would be expected, since as temperature increases cooling costs would also increase.

b A negative correlation would be expected, since as interest rates climb fewer people would be submitting applications for loans.

c A positive correlation would be expected, since husbands and wives tend to have jobs in similar or related classifications. That is, a spouse would be reluctant to take a low-paying job if the other spouse had a high-paying job.

d No correlation would be expected, because those people with a particular I.Q. level would have heights ranging from short to tall.

e A positive correlation would be expected, since people who are taller tend to have larger feet and people who are shorter tend to have smaller feet.

f A weak to moderate positive correlation would be expected. There are some who do well on both, some who do poorly on both, and some who do well on one but not the other. It is perhaps the case that those who score similarly on both tests outnumber those who don't.

g A negative correlation would be expected, since there is a fixed amount of time and as time spent on homework increases, time in watching television would decrease.

h No correlation overall, because for small or substantially large amounts of fertilizer yield would be small.

5.3

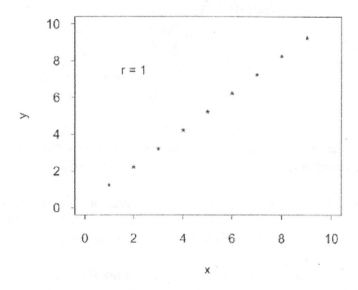

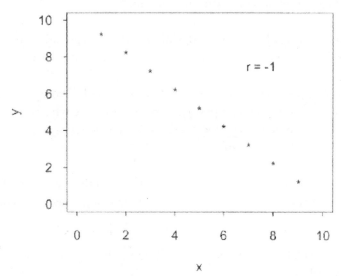

5.5   **a**

Sugar Consumption x	Depression Rate y	z_x	z_y	$z_x z_y$
150	2.3	-1.726	-1.470	2.537
300	3.0	-.369	-.947	.349
350	4.4	.083	.099	.008
375	5.0	.309	.548	.169
390	5.2	.444	.697	.309
480	5.7	1.259	1.071	1.348
				4.720

$$r = \frac{\Sigma z_x z_y}{n-1} = \frac{4.720}{5} = 0.944$$

The correlation is strong and positive.

b   Increasing sugar consumption doesn't cause or lead to higher rates of depression, it may be another reason that causes an increase in both. For instance, a high sugar consumption may indicate a need for comfort food for a reason that also causes depression.

c   These countries may not be representative of any other countries. It may be that only these countries have a strong positive correlation between sugar consumption and depression rate and other countries may have a different type of relationship between these factors. It is therefore not a good idea to generalize these results to other countries.

5.7   The correlation coefficient between household debt and corporate debt would be positive, and the relationship would be strong. As the household debt increases, the corporate debt increases at a similar rate; this can clearly be seen on the graph by a constant width between the two lines.

5.9   **a** $r = 0.335$ There is a weak positive relationship between timber sales and the amount of acres burned in forest fires.

        b   No. Looking at the scatterplot, there doesn't seem much of a relationship at all. Even if there was, it wouldn't necessarily mean one caused the other. In fact out of the 6 years with the highest timber sales, 2 of them had the least amount of acreage lost due to forest fires.

5.11   **a**   Since the points tend to be close to the line, it appears that x and y are strongly correlated in a positive way.

        b   An r value of .9366 indicates a strong positive linear relationship between x and y.

        c   If x and y were perfectly correlated with $r = 1$, then each point would lie exactly on a line. The line would not necessarily have slope 1 and intercept 0.

5.13

$$r = \frac{27918 - \dfrac{(9620)(7436)}{2600}}{\sqrt{36168 - \dfrac{(9620)(9620)}{2600}}\sqrt{23145 - \dfrac{(7436)(7436)}{2600}}}$$

$$= \frac{27918 - 27513.2}{\sqrt{574}\sqrt{1878.04}} = \frac{404.8}{(23.9583)(43.336)} = .3899$$

5.15   No. An r value of −0.085 indicates an extremely weak relationship between support for environmental spending and degree of belief in God.

Exercises 5.17 – 5.31

5.17   **a**

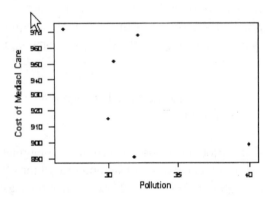

There is a weak negative association between pollution and the cost of medical care.

b   x = Pollution and y = cost; $\Sigma\,x = 191.1, \Sigma\,x^2 = 6184.05\ \Sigma\,xy = 177807,$
$\Sigma\,y = 5597, n = 6,\ \bar{x} = 31.85, \bar{y} = 932.833$

$$S_{xy} = \Sigma\,xy - \frac{(\Sigma\,x)(\Sigma\,y)}{n} = 177807 - \frac{(191.1)(5597)}{6} = -457.45$$

$$S_{xx} = \Sigma\,x^2 - \frac{(\Sigma\,x)^2}{n} = 6184.05 - \frac{(191.1)^2}{6} = 97.515$$

The slope, $b = \dfrac{S_{xy}}{S_{xx}} = \dfrac{-457.45}{97.515} = -4.69$

The intercept, $a = \bar{y} - b\bar{x} = 932.833 - (-4.69)(31.85) = 1082.21$

The equation: $\hat{y} = 1082.21 - 4.69x$

c The slope is negative, consistent with the description in part **a**.

d Yes, it does support the conclusion that elderly people that live in more polluted areas have higher medical costs, but care must be taken not to state that the pollution *causes* the high medical costs – or even the high medical costs causes the pollution!

5.19 **a**

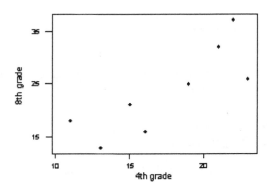

There is a moderately strong positive linear relationship between the percentage of public schools who were at or above the proficient level in math in 4th and 8th grade in the 8 states.

b x = 4th grade and y = 8th grade; $\Sigma x = 140$, $\Sigma x^2 = 2586$ $\Sigma xy = 3497$,
$\Sigma y = 188$, n = 8, $\bar{x} = 17.5, \bar{y} = 23.5$

$$S_{xy} = \Sigma xy - \frac{(\Sigma x)(\Sigma y)}{n} = 3497 - \frac{(140)(188)}{8} = 207$$

$$S_{xx} = \Sigma x^2 - \frac{(\Sigma x)^2}{n} = 2586 - \frac{(140)^2}{8} = 136$$

The slope, $b = \dfrac{S_{xy}}{S_{xx}} = \dfrac{207}{136} = 1.522$

The intercept, $a = \bar{y} - b\bar{x} = 23.5 - 1.522(17.5) = -3.135$

The equation: $\hat{y}$ = -3.135 + 1.522x

c Predicted 8th grade = -3.135 + 1.522(4th grade percent) $\Rightarrow$ -3.135 + 1.522(14) = 18 (rounded to nearest integer). This is 2% lower than the actual 8th grade value of 20 for Nevada.

5.21   **a**

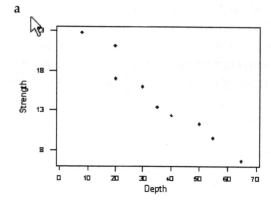

There appears to be a negative linear association between carbonation depth and the strength of concrete for a sample of core specimens.

b $\Sigma x = 323,\ \Sigma x^2 = 14339,\ \Sigma xy = 3939.9,$
$\Sigma y = 130.8,\ n = 9,\ \bar{x} = 35.889,\ \bar{y} = 14.533$

$$S_{xy} = \Sigma xy - \frac{(\Sigma x)(\Sigma y)}{n} = 3939.9 - \frac{(323)(130.8)}{9} = -754.367$$

$$S_{xx} = \Sigma x^2 - \frac{(\Sigma x)^2}{n} = 14339 - \frac{(323)^2}{9} = 2746.889$$

The slope, $b = \dfrac{S_{xy}}{S_{xx}} = \dfrac{-754.367}{2746.889} = -0.275$

The intercept, $a = \bar{y} - b\bar{x} = 14.533 - (-0.275)(35.889) = 24.40$

The equation: $\hat{y} = 24.4 - 0.275x$

c   When depth is 25, predicted strength $= 24.4 - 0.275(25) = 17.5$

d   The least squares line was calculated using values of "depth" of between 8 mm and 65 mm and therefore is only valid for values in this range. We don't know if the relationship between depth and strength remains the same outside these values and so this equation cannot be used. A depth of 100 mm is clearly outside these values and it would be unreasonable to use this equation to predict strength.

5.23   The slope is the average increase in the y variable for an increase of one unit in the x variable. Because the home prices (y variable) dropped by an average of $4000 (-4000) for every (1) mile (x variable) from the Bay area, the slope is -4000/1 = -4000.

5.25   **a** $\Sigma x = 240,\ \Sigma x^2 = 6750\ \Sigma xy = 199750,$
$\Sigma y = 7250,\ n = 11,\ \bar{x} = 21.818, \bar{y} = 659.091$

$$S_{xy} = \sum xy - \frac{(\sum x)(\sum y)}{n} = 199750 - \frac{(240)(7250)}{11} = 41568.182$$

$$S_{xx} = \sum x^2 - \frac{(\sum x)^2}{n} = 6750 - \frac{(240)^2}{11} = 1513.636$$

The slope, $b = \dfrac{S_{xy}}{S_{xx}} = \dfrac{41568.182}{1513.636} = 27.462$

The intercept, $a = \bar{y} - b\bar{x} = 659.091 - 27.462(21.818) = 59.925$

The equation: $\hat{y} = 59.925 + 27.462x$

b Concentration with 18% bare ground: 59.925 + 27.462(18) = 554 (to nearest integer)

c No, because the data used to obtain the least squares equation was from steeply sloped plots, so it would not make sense to use it to predict runoff sediment from gradually sloped plots. You would need to use data from gradually sloped plots to create a least squares regression equation to predict runoff sediment from gradually sloped plots.

5.27 **a**

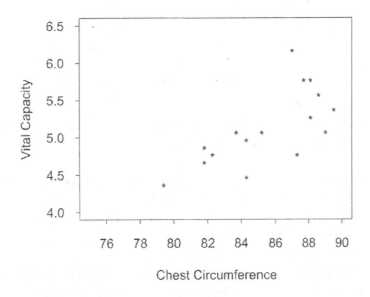

The graph reveals a moderate linear relationship between x and y.

b

$$b = \frac{6933.48 - \dfrac{(1368.1)(80.9)}{16}}{117123.85 - \dfrac{(1368.1)^2}{16}} = \frac{6933.48 - 6917.456}{117123.85 - 116981.101} = \frac{16.0244}{142.7494} = 0.1123$$

$$a = \frac{80.9}{16} - 0.1123\left(\frac{1368.1}{16}\right) = 5.0563 - 0.1123(85.5063) = 5.0563 - 9.6024 = -4.5461$$

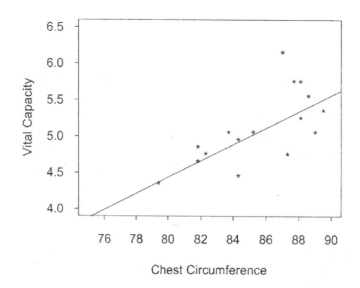

Chest Circumference

c   The change in vital capacity associated with a 1 cm. increase in chest circumference is .1123.

The change in vital capacity associated with a 10 cm. increase in chest circumference is 10(.1123) = 1.123.

d $\hat{y} = -4.5461 + .1123(85) = 4.9994$

e   No; this is shown by the fact that there are two data points in the data set whose x values are 81.8, but these data points have different y values.

5.29   **a**   y = 100 + .75(s$_y$)2 = 100 + 1.5(s$_y$). That person's annual sales would be 1.5 standard deviations above the mean sales of 100.

b $(y-\bar{y}) = r\dfrac{s_y}{s_x}(x-\bar{x})$, which implies $\dfrac{y-\bar{y}}{s_y} = r\dfrac{x-\bar{x}}{s_x}$. Hence, $-1.0 = r(-1.5)$ implies

$$r = \dfrac{-1.0}{-1.5} = .67.$$

5.31   **a** $\hat{y} = -424.7 + 3.891x$

        b   Let $y' = cy$. Then $\bar{y}' = c\bar{y}$.

$$b' = \dfrac{\sum(x-\bar{x})(cy-c\bar{y})}{\sum(x-\bar{x})^2} = \dfrac{c\sum(x-\bar{x})(y-\bar{y})}{\sum(x-\bar{x})^2} = cb$$

$$a' = c\bar{y} - cb\bar{x} = c(\bar{y} - b\bar{x}) = ca$$

        Both the slope and the y intercept are changed by the multiplicative factor c. Thus, the new least squares line is the original least squares line multiplied by c.

Exercises 5.32 – 5.46

5.33   **a**

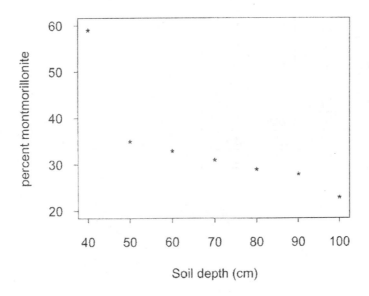

b

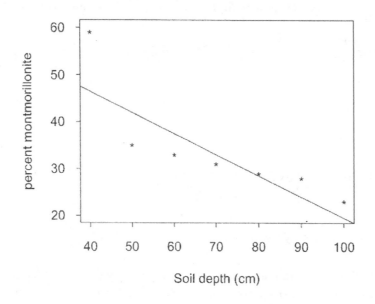

Yes, there appear to be large residuals, those associated with the x-values of 40, 50 and 60.

c

x	y	$\hat{y}$	y - $\hat{y}$
40	58	46.5	11.5
50	34	42.0	−8.0
60	32	37.5	−5.5
70	30	33.0	−3.0
80	28	28.5	−0.5
90	27	24.0	3.0
100	22	19.5	2.5

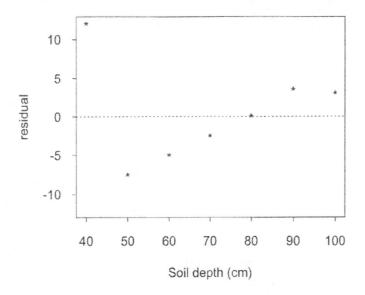

Soil depth (cm)

Yes, the residuals for small x-values and large x-values are positive, while the residuals for the middle x-values are negative.

5.35 From problem 5.60, the equation of the least-squares line is $\hat{y} = 94.33 - 15.388x$.

x	y	$\hat{y}$	residual
.106	98	92.6989	5.30112
.193	95	91.3601	3.63988
.511	87	86.4667	0.53326
.527	85	86.2205	-1.22053
1.08	75	77.7110	-2.71096
1.62	72	69.4014	2.59856
1.73	64	67.7088	-3.70876
2.36	55	58.0143	-3.01432
2.72	44	52.4746	-8.47464
3.12	41	46.3194	-5.31945
3.88	37	34.6246	2.37544
4.18	40	30.0082	9.99184

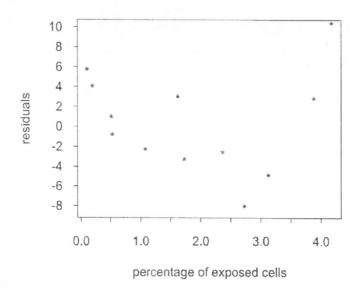

There appears to be a pattern in the plot. It is like the graph of a quadratic equation.

5.37   **a** $r^2 = 15.4\%$

b $r^2 = 16\%$: No, only 16% of the variability in first-year grades can be attributed to an approximate linear relationship between first-year college grades and SAT II score so this does not indicate a good predictor.

5.39   **a**   There does appear to be a positive linear relationship between x and y.

b $\sum x = 798 \quad \sum x^2 = 63040 \quad \sum y = 643 \quad \sum y^2 = 41999 \quad \sum xy = 51232$

$\bar{x} = \dfrac{798}{15} = 53.2 \quad \bar{y} = \dfrac{643}{15} = 42.87$

$b = \dfrac{51232 - \dfrac{(798)(643)}{15}}{63040 - \dfrac{(798)(798)}{15}} = \dfrac{17024.4}{20586.4} = 0.827$

$a = 42.87 - 0.827(53.2) = -1.13$

$\hat{y} = -1.13 + 0.827x$

c   For x=80, $\hat{y} = -1.13 + 0.827(80) = 65.03$

d $SSResid = \sum y^2 - a\sum y - b\sum xy$

$\qquad\qquad = 41999 - (-1.13)(643) - 0.827(51232)$

$\qquad\qquad = 356.726$

$$s_e = \sqrt{\frac{356.726}{15-2}} = 5.238$$

e $r^2 = 1 - \dfrac{SSresid}{SSTo}$

$SSTo = \sum y^2 - \dfrac{(\sum y)^2}{n} = 41999 - \dfrac{(643)^2}{15} = 14435.734$

$r^2 = 1 - \dfrac{356.726}{14435.734} = 1 - 0.025 = 0.975$

So 97.5% of the observed variation in runoff volume can be explained by the linear relationship to rainfall volume.

5.41   **a** $\hat{y} = -89.09 + .72907(375) = 184.31$

residual $= 165 - 184.31 = -19.31$

   b $r^2 = .963$

5.43 $r^2 = 1 - \dfrac{1235.470}{25321.368} = 0.9512$

The coefficient of determination reveals that 95.12% of the total variation in hardness of molded plastic can be explained by the linear relationship between hardness and the amount of time elapsed since termination of the molding process.

5.45   **a**   Whether s_e is small or not depends upon the physical setting of the problem. An s_e of 2 feet when measuring heights of people would be intolerable, while an s_e of 2 feet when measuring distances between planets would be very satisfactory. It is possible for the linear association between x and y to be such that r^2 is large and yet have a value of s_e that would be considered large. Consider the following two data sets:

	Set 1			Set 2	
	x	y		x	y
	5	14		14	5
	6	16		16	15
	7	17		17	25
	8	18		18	35
	9	19		19	45
	10	21		21	55

For set 1, $r^2 = .981$ and $s_e = .378$. For set 2, $r^2 = .981$ and $s_e = 2.911$.
Both sets have a large value for r^2, but s_e for data set 2 is 7.7 times larger than s_e for data set 1. Hence, it can be argued that data set 2 has a large r^2 and a large s_e.

b   Now consider the data set

x	5	55	15	45	25	35
y	10.004	10.006	10.007	10.008	10.009	10.010

This data set has $r^2 = .12$ and $s_e = .002266$. So yes, it is possible for a bivariate data set to have both r^2 and s_e small.

c   When r^2 is large and s_e is small, then not only has a large proportion of the total variability in y been explained by the linear association between x and y, but the typical error of prediction is small.

Exercises 5.47 – 5.54

5.47   **a**

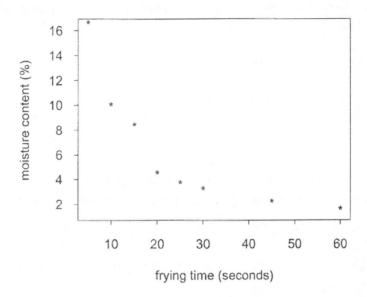

Because of the substantial curvature in the plot, a straight line would not provide an effective summary of the relationship.

      b

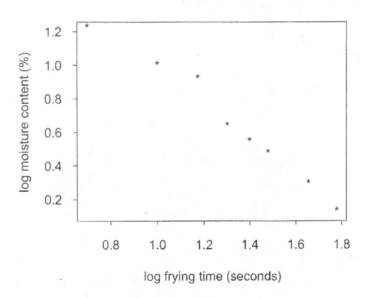

The plot of the transformed variables suggests that it could be modeled by a straight line.

c   The coefficient of determination between y' and x' is .976. This suggests that a least-squares line might effectively summarize the relationship between x' and y'.

d   When $x = 35, x' = 1.54407$

$\hat{y}' = 2.01444 - 1.0492(1.54407) = .394404$

$\hat{y} = 10^{.394404} = 2.47973$

e   Yes, this appears to be the case. To see this, predict y using both approaches. Compare the values of $\sum (y - \hat{y})^2$ for the two methods. The method of part **c** results in a lower value for $\sum (y - \hat{y})^2$.

5.49   a $n = 12 \quad \sum x = 22.4 \quad \sum y = 303.1 \quad \sum x^2 = 88.58 \quad \sum y^2 = 12039.27$
$\sum xy = 241.29$

$$r = \frac{241.29 - \dfrac{(22.4)(303.1)}{12}}{\sqrt{88.58 - \dfrac{(22.4)^2}{12}}\sqrt{12039.27 - \dfrac{(303.1)^2}{12}}} = \frac{-324.50}{\sqrt{46.767}\sqrt{4383.47}}$$

$$= \frac{-324.5}{(6.84)(66.208)} = -0.717$$

b $n = 12 \quad \sum x = 13.5 \quad \sum y = 55.74 \quad \sum x^2 = 22.441 \quad \sum y^2 = 303.3626$
$\sum xy = 47.7283$

$$r = \frac{47.7283 - \dfrac{(13.5)(55.74)}{12}}{\sqrt{22.441 - \dfrac{(13.5)^2}{12}}\sqrt{303.3626 - \dfrac{(55.74)^2}{12}}} = \frac{-14.9792}{\sqrt{7.2535}\sqrt{44.4503}}$$

$$\frac{-14.9792}{(2.693)(6.667)} = -0.835$$

The correlation between $\sqrt{x}$ and $\sqrt{y}$ is $-.835$. Since this correlation is larger in absolute value than the correlation of part **a**, the transformation appears successful in straightening the plot.

5.51   **a**

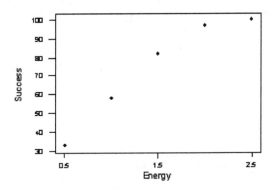

The relationship appears non-linear.

b $\Sigma\, x = 7.5,\ \Sigma\, x^2 = 13.75\ \ \Sigma\, xy = 641.05,$
$\Sigma\, y = 370.1,\ n = 5,\ \ \bar{x} = 1.5, \bar{y} = 74.02$

$$S_{xy} = \Sigma\, xy - \frac{(\Sigma\, x)(\Sigma\, y)}{n} = 641.05 - \frac{(7.5)(370.1)}{5} = 85.9$$

$$S_{xx} = \Sigma\, x^2 - \frac{(\Sigma\, x)^2}{n} = 13.75 - \frac{(7.5)^2}{5} = 2.5$$

The slope, $b = \dfrac{S_{xy}}{S_{xx}}\ \dfrac{85.9}{2.5} = 34.36$

The intercept, $a = \bar{y} - b\bar{x} = 74.02 - 34.36(1.5) = 22.48$

The equation: $\hat{y} = 22.48 + 34.36x$

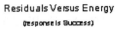
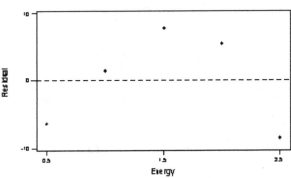

There is a definite curvature in the residual plot confirming the conclusion in part **a**.

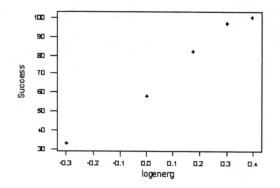

Residuals Versus logenerg

(response is Success)

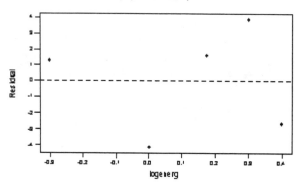

The value of r² is higher and the size of the residuals are smaller for the log transformation.

d $y = a + b(x')$ where $x' = \log_{10}(x)$

values of x' are: -0.30102, 0, 0.17609, 0.30103, 0.39794

$\Sigma x' = .5740$, $\Sigma (x')^2 = .3706$ $\Sigma x'y = 73.2836$,

$\Sigma y = 370.1$, $n = 5$, $\bar{x}' = .1148$, $\bar{y} = 74.02$

$$S_{xy} = \Sigma xy - \frac{(\Sigma x)(\Sigma y)}{n} = 73.2836 - \frac{(.5740)(370.1)}{5} = 30.796$$

$$S_{xx} = \Sigma x^2 - \frac{(\Sigma x)^2}{n} = .3706 - \frac{(.574)^2}{5} = 0.3047$$

The slope, $b = \dfrac{S_{xy}}{S_{xx}} = \dfrac{30.796}{0.3047} = 101.07$

The intercept, $a = \bar{y} - b\bar{x} = 74.02 - 101.07(.1148) = 62.417$

The equation: $\hat{y} = 62.417 + 101.07x' \implies \hat{y} = 62.417 + 101.07 \log(x)$

e When energy of shock (x) = 1.75, predicted success percent to be 62.417 + 101.07(log 1.75) = 87.0%. When energy of shock is 0.8, the predicted success would be 62.417 + 101.07(log 0.8) = 52.6%

5.53   **a**

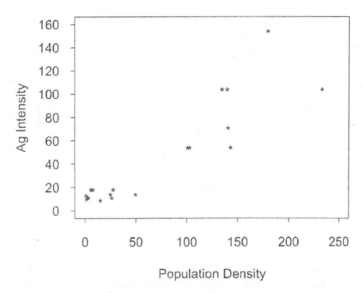

Population Density

The plot does appear to have a positive slope, so the scatter plot is compatible with the "positive association" statement made in the paper.

b

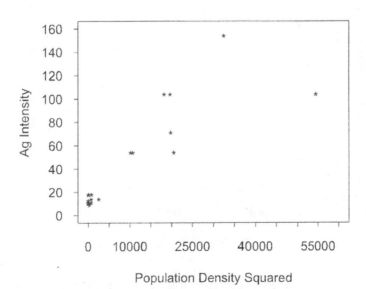

Population Density Squared

This transformation does straighten the plot, but it also appears that the variability of y increases as x increases.

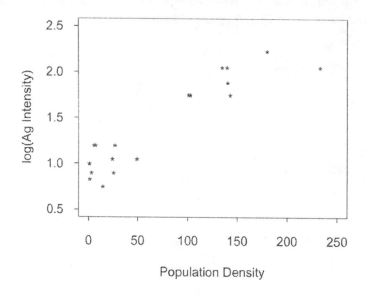

The plot appears to be as straight as the plot in **b**, and has the desirable property that the variability in y appears to be constant regardless of the value of x.

d

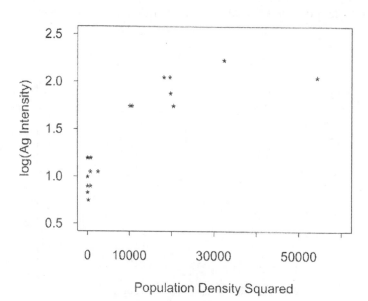

This plot has curvature opposite of the plot in **a**, suggesting that this transformation has taken us too far along the ladder.

Exercises 5.55 – 5.67

5.55 **a** $r = -0.981$ There appears to be a strong negative linear relationship between the amount of catalyst added to a chemical reaction and the resulting reaction time.

b

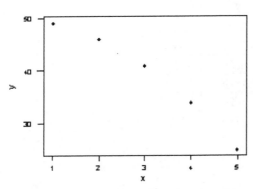

There is a definite curvature to the plot. Linear does not seem the best description of this relationship. This shows the importance of checking, not only the numerical checks of "a good fit" but also graphical ones too.

5.57 **a** $n = 15$, $\sum x = 82.82$, $\sum y = 12545$, $\sum x^2 = 459.9784$, $\sum y^2 = 12734425$, $\sum xy = 67703.9$

$$\sum xy - \frac{(\sum x)(\sum y)}{n} = 67703.9 - \frac{(82.82)(12545)}{15} = -1561.227$$

$$\sum x^2 - \frac{(\sum x)^2}{n} = 459.9784 - \frac{(82.82)^2}{15} = 2.702$$

$$b = \frac{-1561.227}{2.702} = -577.895$$

$$a = 836.33 - (-577.8953)(5.5213) = 4027.083$$

$$\hat{y} = 4027.083 - 577.895x$$

b The b value of –577.895 is the estimate of the average change in myoglobin level associated with a one unit increase in finishing time.

c $\hat{y} = 4027.083 - 577.895(8) = -596.077$

The least squares equation yields a negative value for the estimated level of myoglobin when the finishing time is 8h. This is clearly unreasonable since myoglobin level cannot be a negative value.

5.59   **a**   The least square line is $\hat{y} = 32.08 + 0.5549x$

x	y	predicted	residual
15	23	40.4048	−17.4048
19	52	42.6245	9.3755
31	65	49.2837	15.7163
39	55	53.7231	1.2769
41	32	54.8330	−22.8330
44	60	56.4978	3.5022
47	78	58.1626	19.8374
48	59	58.7175	0.2825
55	61	62.6020	−1.6020
65	60	68.1513	−8.1513

b
$$\text{SSResid} = (-17.4048)^2 + (9.3755)^2 + \ldots + (-1.6020)^2 + (-8.1513)^2$$
$$= 302.9271 + 87.9000 + \ldots + 2.5664 + 66.4437$$
$$= 1635.6833$$

$$\text{SSTo} = 31993 - \frac{(545)^2}{10} = 31993 - 29702.5 = 2290.50$$

$$r^2 = 1 - \frac{1635.6833}{2290.5000} = 1 - .7141 = .2859$$

c   The least squares line does not give very accurate predictions. Only 28.59% of the observed variation in age is explained by the linear relationship between percent of root transparent dentine and age.

5.61   **a**

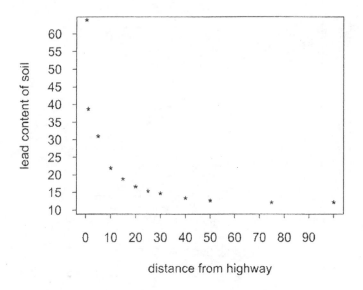

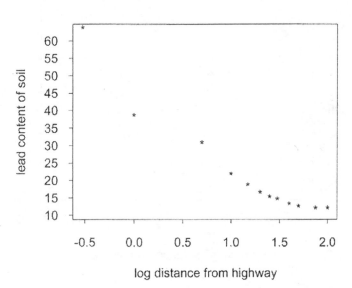

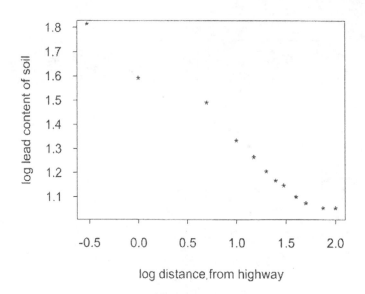

b   It appears that log x, log y does the best job of producing an approximate linear relationship. The least-squares equation for predicting $y' = \log y$ from $x' = \log x$ is $\hat{y}' = 1.61867 - .31646x'$.

When $x = 25, x' = 1.39764$

$\hat{y}' = 1.61867 - .31646(1.39764) = 1.17628$

$\hat{y} = 10^{1.17628} = 15.0064$

82

5.63 **a** The plot does not suggest a linear relationship. However, the one outlier value (51.3, 49.3) prevents an accurate interpretation.

b $\hat{y} = -11.37 + 1.0906(40) = 32.254$

c About 59.5 percent of the variability in fire-simulation consumption is explained by the linear relationship between treadmill consumption and fire-simulation consumption.

d For the new data set, $n = 9$, $\sum x = 388.8 - 51.3 = 337.5$, $\sum y = 310.3 - 49.3 = 261.0$

$\sum x^2 = 15338.54 - (51.3)^2 = 12706.85$, $\sum y^2 = 10072.41 - (49.3)^2 = 7641.92$

$\sum xy = 12306.58 - (51.3)(49.3) = 9777.49$

$$\sum x^2 - \frac{\left(\sum x\right)^2}{n} = 12706.85 - \frac{(337.5)^2}{9} = 50.60$$

$$\sum xy - \frac{\left(\sum x\right)\left(\sum y\right)}{n} = 9777.49 - \frac{(337.5)(261.0)}{9} = -10.01$$

$$b = \frac{-10.01}{50.60} = -0.1978, \quad a = \frac{261}{9} - (-.1978)\left(\frac{337.5}{9}\right) = 36.4175$$

$\hat{y} = 36.4175 - .1978x$

$$\sum y^2 - \frac{\left(\sum y\right)^2}{n} = 7641.92 - \frac{(261)^2}{9} = 72.92, \quad r^2 = \frac{(-10.01)^2}{(50.6)(72.92)} = .027$$

Without the observation (51.3, 49.3) there is very little evidence of a linear relationship between fire-simulation consumption and treadmill consumption. One would be very hesitant to use the prediction equation based on the data set including this observation because this observation is very influential.

5.65 **a** $\sum(x - \bar{x})^2 = 2, \quad \sum(y - \bar{y})^2 = 2, \quad \sum(x - \bar{x})(y - \bar{y}) = 0$

$$r = \frac{0}{\sqrt{2(2)}} = 0$$

b If $y = 1$, when $x = 6$, then $r = .509$.
(Comment: Any y value greater than .973 will work.)

83

If y = –1, when x = 6, then r = –.509.

(Comment: any y value less than –.973 will work).

5.67   **a**

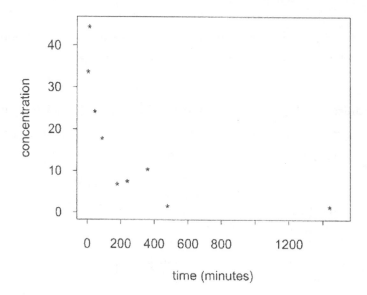

b   Based on the plot in **a** and figure 5.34 a transformation going down the ladder on x or y is suggested. The transformation log(time) will produce a reasonably straight plot.

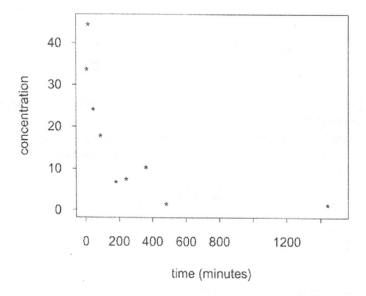

Devore & Peck   Chapter 6

Exercises 6.1 – 6.14

6.1 **a** In the long run, 1% of all people who suffer cardiac arrest in New York City survive.

b 1% of 2329 is .01(2329) = 23.29. So in this study there must have been only 23 or 24 survivors.

6.3 A randomly selected adult is more likely to experience pain daily if they are a woman. The events are dependent.

6.5 The outcomes *selected student is a recent immigrant* and *selected student has TB* are dependent events. Knowing that the selected student is a recent immigrant increases the probability of having TB from .0006 to .0075.

6.7 The outcome *'selected smoker who is trying to quit uses a nicotine aid'* and *'selected smoker who had attempted to quit begins smoking again within two weeks'* are dependent events. Knowing that the selected smoker used a nicotine aid reduces the probability of beginning smoking again within 2 weeks from 0.62 to 0.60.

6.9 **a** Assuming the events are independent,

P(Jeanie forgets all three errands)

= P(Jeanie forgets first) x P(Jeanie forgets second) x P(Jeanie forgets third)

= (0.1)(0.1)(0.1) = 0.001

b P(Jeanie remembers at least one)

= 1 – P(Jeanie forgets all three) = 1 – .001 = .999

c P(Jeanie remembers the first errand but not the second or third)

= P(Jeanie remembers first) x P(Jeanie forgets second) x P(Jeanie forgets third)

= (1 – 0.1)(0.1)(0.1) = (0.9)(0.1)(0.1) = 0.009

6.11 **a** The expert assumed that the positions of the two valves were independent.

b If the car was driven in a straight line, the relative positions of the valves would remain unchanged. Thus, the probability of them ending up at a one o'clock, six o'clock position would be 1/12, not 1/144. The value 1/144 is smaller than the correct probability of occurrence.

6.13 **a** P(1–2 subsystem works) = P(1 works) · P(2 works) = (.9)(.9) = .81

  b P(1–2 subsystem doesn't work) = 1 – P(1–2 subsystem works) = 1 – .81 = .19

   P(3–4 subsystem doesn't work) = P(1–2 subsystem doesn't work) = .19

  c P(system won't work)

   = P(1–2 subsystem doesn't work) x P(3–4 subsystem doesn't work)

   = (.19)(.19) = .0361

   P(system will work) = 1 – .0361 = .9639

  d P(system won't work) = (.19)(.19)(.19) = .006859

   P(system will work) = 1 – .006859 = .993141

  e P(1–2 subsystem works) = (.9)(.9)(.9) = .729

   P(1–2 subsystem won't work) = 1 – .729 = .271

   P(system won't work) = (.271)(.271) = .073441

   P(system works) = 1 – .073441 = .926559

Exercises 6.15 – 6.18

6.15 Reliability of predicting female babies = P(baby is born female given that she is predicted to be female)= $\dfrac{Total\ number\ of\ predicted\ female\ babies\ born\ female}{Total\ number\ of\ predicted\ female\ babies} = \dfrac{432}{562} = 0.769$

Reliability of predicting male babies = P(baby is born male given that he is predicted to be male)= $\dfrac{Total\ number\ of\ predicted\ male\ babies\ born\ male}{Total\ number\ of\ predicted\ male\ babies} = \dfrac{390}{438} = 0.890$

Since the reliability of predicting male babies is higher than for female babies, it would appear that ultrasound is more reliable in predicting boys.

6.17 There are a total of 96 current smokers. 60 of them view them view smoking as very harmful. There are a total of 99 former smokers and 99 who have never smoked. 78 former smokers and 86 non-smokers view smokers as very harmful. As this data can be considered representative of the U.S. adult population, the probability of a current smoker viewing smoking as very harmful is

$\dfrac{60}{96} = 0.625$ compared to $\dfrac{78}{99} = 0.7879$ for a former smoker and $\dfrac{86}{99} = 0.8687$ for someone who has never smoked, so the conclusion is justified.

Exercises 6.19 – 6.27

6.19 a $\dfrac{425}{500} = .85$

  b $\dfrac{500 - 405}{500} = \dfrac{95}{500} = .19$

  c $\left(\dfrac{415}{500}\right)\left(\dfrac{415}{500}\right) = (.83)(.83) = .6889$

  d $\dfrac{5220}{6000} = .87$

6.21 a   P (selected student is a male) = $\dfrac{6700}{17000} = 0.3941$

  b   P (selected student is in the College of Agriculture) = $\dfrac{3000}{17000} = 0.1765$

  c   P (selected student is male and is in the College of Agriculture) = $\dfrac{900}{17000} = 0.0529$

  d   P (selected student is male and not from Agriculture) = $\dfrac{5800}{17000} = 0.3412$

6.23 The results will vary from one simulation to another. The approximate probabilities given were obtained from a simulation done by computer with 100,000 trials.

  a   .71293

  b   .02201

6.25   The simulation results will vary from one simulation to another. The approximate probability should be around .8504.

6.27   **a**   The simulation results will vary from one simulation to another. The approximate probability should be around .6504.

b   The decrease in the probability of on time completion for Jacob made the biggest change in the probability that the project is completed on time.

Exercises 6.28 – 6.34

6.29   The total number of observations is

20 + 72 + 209 + ... + 393 + 633 = 19380.

a   P(visiting team got fewer than 3 hits) $= \dfrac{20 + 72 + 209}{19380} = \dfrac{301}{19380} = .0155$

b   P(home team got more than 13 hits) $= \dfrac{569 + 393 + 633}{19380} = \dfrac{1595}{19380} = .0823$

c   P(a team gets 10 or more hits)

$$= \frac{1967 + 1509 + 1230 + 843 + 569 + 393 + 633}{19380} = \frac{7144}{19380} = .3686$$

P(both teams got 10 or more hits)

= P(visiting team got 10 or more hits) x P(home team got 10 or more hits)

$$= \left(\frac{7144}{19380} \right)\left(\frac{7144}{19380} \right) = (0.3686)(0.3686) = 0.1359$$

d   I know of no reason to believe that knowing that one team had 10 or more hits would change the probability that the other team would have 10 or more hits. So I think the independence assumption is reasonable.

Supplementary Exercises

6.31   **a**   P(request is for one of the 3 B's) = .05 + .26 + .09 = .40

b   P(not for one of the two S's) = 1 – P(for one of the two S's) = 1 – (.12 + .07) = .81

c P(request for a composer who wrote at least one symphony)

    = 1 − P(composer did not write a symphony)

    = 1 − (.05 + .01) = .94

6.33 **a** $(.3)^5 = .00243$

  b $(.3)^5 + (.2)^5 = .00243 + .00032 = .00275$

  c Select a random digit. If it is a 0, 1, or 2, then award A one point. If it is a 3 or 4, then award B one point. If it is 5, 6, 7, 8, or 9, then award A and B one-half point each. Repeat the selection process until a winner is determined or the championship ends in a draw (5 points each). Replicate this entire procedure a large number of times. The estimate of the probability that A wins the championship would be the ratio of the number of times A wins to the total number of replications.

  d It would take longer if no points for a draw are awarded. It is possible that a number of games would end in a draw and so more games would have to be played in order for a player to earn 5 points.

6.35 **a** Choose a random digit. If it is a 1, 2, 3, 4, 5, 6, 7 or 8, then seed 1 defeats seed 4. If it is a 9 or 0, then seed 4 defeats seed 1.

  b Choose a random digit. If it is a 1, 2, 3, 4, 5 or 6, then seed 2 defeats seed 3. If it is a 7, 8, 9, or 0, then seed 3 defeats seed 2.

  c If seeds 1 and 2 have won in the first round, then they are competing in game 3. Select a random digit. If it is a 1, 2, 3, 4, 5 or 6, then seed 1 defeats seed 2 in game 3. If it is a 7, 8, 9, or 0, then seed 2 defeats seed 1 in game 3.

    If seeds 1 and 3 have won in the first round, then they are competing in game 3. Select a random digit. If it is a 1, 2, 3, 4, 5, 6, or 7, then seed 1 defeats seed 3 in game 3. If it is an 8, 9, or 0, then seed 3 defeats seed 1 in game 3.

    If seeds 2 and 4 have won in the first round, then they are competing in game 3. Select a random digit. If it is a 1, 2, 3, 4, 5, 6, or 7, then seed 2 defeats seed 4 in game 3. If it is an 8, 9, or 0, then seed 4 defeats seed 2 in game 3.

    If seeds 3 and 4 have won in the first round, then they are competing in game 3.

Select a random digit. If it is a 1, 2, 3, 4, 5, or 6, then seed 3 defeats seed 4 in game 3. If it is a 7, 8, 9, or 0, then seed 4 defeats seed 3 in game 3.

d   For game 1, selected digit is 7. Seed 1 wins game 1.
For game 2, selected digit is 9. Seed 3 wins game 2.
So seeds 1 and 3 compete in game 3.
For game 3, selected digit is 6. Seed 1 wins game 3 and the tournament.

e   Summarize the result of part **d** as follows: digits selected $(7,9,6) \rightarrow$ seed 1 wins tournament.

Replications	Digits Selected	Winner of Tournament
2	(5,2,9)	seed 2
3	(0,6,0)	seed 4
4	(7,5,8)	seed 2
5	(0,4,1)	seed 2
6	(6,6,9)	seed 2
7	(3,1,2)	seed 1
8	(2,6,2)	seed 1
9	(7,5,0)	seed 2
10	(5,6,4)	seed 1

Based on this simulation of 10 trials, the estimate of the probability that the first seed wins the tournament is 4/10 = .4.

f   Answers will vary.

g   The answers for parts **e** and **f** differ because they are based on different random selections and different number of trials. Generally, the larger the number of trials, the better the estimate is. So one would believe that the estimate from part **f** is better than the estimate of part **e**.

Exercises 6.1 – 6.13

6.1   A chance experiment is any activity or situation in which there is uncertainty about which of two or more possible outcomes will result.

Consider tossing a coin two times and observing the outcome of the two tosses. The possible outcomes are (H, H), (H, T), (T, H), and (T, T), where (H, H) means both tosses resulted in Heads, (H, T) means the first toss resulted in a Head and the second toss resulted in a Tail, etc. This is an example of a chance experiment with four possible outcomes.

6.3   a   Sample space = {(A, A), (A, M), (M, A), (M, M)}.

       b

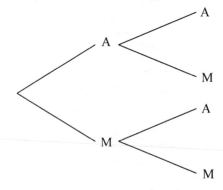

       c   B = {(A, A), (A, M), (M, A)}
             C = {(A, M), (M, A)}
             D = {(M, M)}
             Only D is a simple event.

       d   B and C = {(A, M), (M, A)} = C
             B or C = {(A, A), (A, M), (M, A)} = B

6.5 **a**   # of defective tires   # of defective headlights

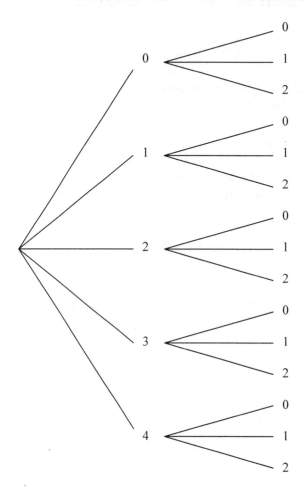

b A^c = {(0,2), (1,2), (2,2), (3,2), (4,2)} where (i,j) means that the number of defective tires is i and the number of defective headlights is j.

A ∪ B = {(0,0), (1,0), (2,0), (3,0), (4,0), (0,1), (1,1), (2,1), (3,1), (4,1), (0,2), (1,2)} where (i,j) means that the number of defective tires is i and the number of defective headlights is j.

A ∩ B = {(0,0), (1,0), (0,1), (1,1)}

c   C = {(4,0), (4,1), (4,2)}. A and C are not disjoint because the outcomes (4,0) and (4,1) are in both A and C. B and C are disjoint because the outcomes in B are of the form (i,j) where i =0 or 1 but the outcomes in C are of the form (i,j) where i=4. Recall that the notation (i,j) means that the number of defective tires is i and the number of defective headlights is j.

6.7   **a**

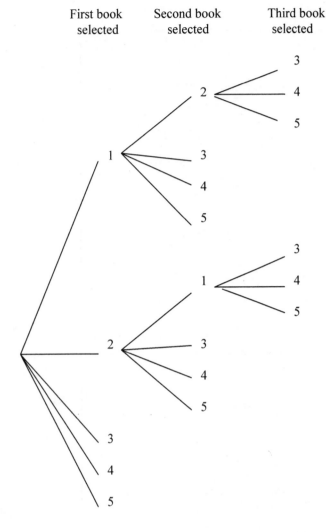

First book selected   Second book selected   Third book selected

b   Exactly one book is examined when the first selected book is copy 3, 4, or 5. So A = {3,4,5}.

c   C = {5, (1,5), (2,5), (1,2,5), (2,1,5)}. The notation used is as follows. The outcome 5 means that copy 5 is selected first. The outcome (1,5) means that copy 1 is selected first and copy 5 is selected second, and so on.

6.9   **a**   A = {NN, DNN, NDN}

b   B = {DDNN, DNDN, NDDN}

c   The number of possible outcomes is infinite.

6.11 **a**

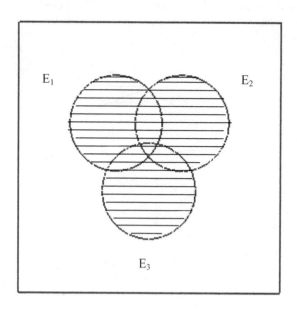

b The required region is the area common to all three circles.

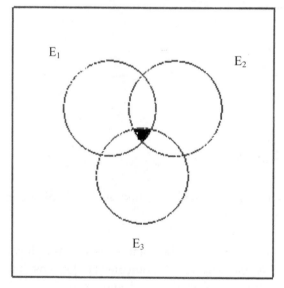

c

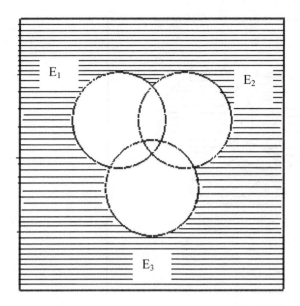

d

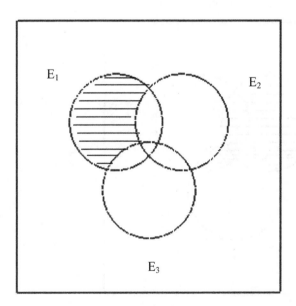

95

e

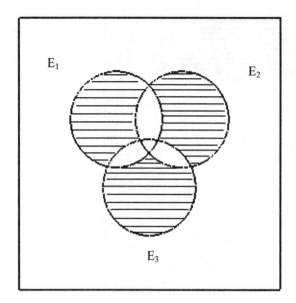

f

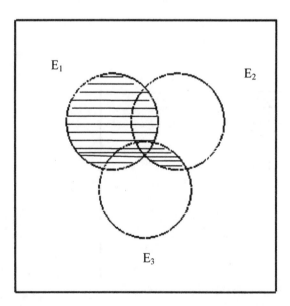

6.13 **a**   Sample space = {n | n is an integer between 0 and 1000 inclusive} =
{0,1,2,........,1000}

       b $E = \{0,1,2,3,4,5\}$; $G = \{6,7,8,9,10\}$; $F = \{11,12,13,14,15,16,17,18,19,20\}$.

       c   Yes, I assigned the outcomes to the events E, G, F so that these events are disjoint.
To remove the ambiguity in the meaning of the word "between", we might

define the rating "Good" to mean "greater than 5 errors but less than or equal to 10 errors per 1000 keystrokes" and the rating "Fair" to mean "greater than 10 errors but less than or equal to 20 errors per 1000 keystrokes".

d

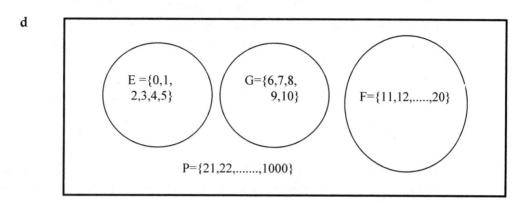

E ={0,1, 2,3,4,5}   G={6,7,8, 9,10}   F={11,12,.....,20}

P={21,22,........,1000}

Exercises 6.14 – 6.28

6.15 **a** In the long run, 1% of all people who suffer cardiac arrest in New York City survive.

b 1% of 2329 is .01(2329) = 23.29. So in this study there must have been only 23 or 24 survivors.

6.17 **a** A = {(C,N), (N,C), (N,N)}. So P(A) = 0.09 + 0.09 + 0.01 = 0.19.

b B = {(C,C), (N,N)}. So P(B) = 0.81 + 0.01 = 0.82.

6.19 **a** In the long run, 35% of all customers who purchase a tennis racket at this store will have a grip size equal to 4 ½ inches.

b P(grip size is not 4 ½ inches) = 1 ◎ 0.35 = 0.65.

c P(oversize head) = 0.20 + 0.15 + 0.20 = 0.55. In the long run, 55% of all customers who purchase a racket at this store will buy a racket with an oversize head.

d P(grip size is at least 4 ½ inches) = 0.20 + 0.15 + 0.15 + 0.20 = 0.70.

6.21 **a** P(request is for one of the three B's) = 0.05 + 0.26 + 0.09 = 0.40.

b P(request is not for one of the two S's) = 1- P(request is for one of the two S's)
= 1-(0.12+0.07) = 0.81.

c P(request is for a composer who wrote at least one symphony)

= 1-P(request is not for Bach or Wagner) = 1-(0.05 + 0.01) = 0.94.

6.23   **a**   P(hand will consist entirely of spades) = $\dfrac{1287}{2598960} = 0.000495198$.

P(hand will consist entirely of a single suit) = P(all spades) + P(all clubs) + P(all diamonds) + P(all hearts) = 0.000495198 + 0.000495198 + 0.000495198 + 0.000495198 = 0.001980792.

   b   P(hand consists entirely of spades and clubs with both suits represented)
$= \dfrac{63206}{2598960} = 0.0243197$.

   c   P(hand contains cards from exactly two suits) = P(spades and clubs) + P(spades and diamonds) + P(spades and hearts) + P(clubs and diamonds) + P(clubs and hearts) + P(diamonds and hearts) = 6(0.0243197) = 0.1459182.

6.25   **a**   The simple events in this experiment are
(C, D, P), (C, P, D), (D, C, P), (D, P, C), (P, C, D), (P, D, C).
All six outcomes may be considered equally likely so the probability assigned to any one outcome is 1/6.

   b   P(C is ranked first) = 2/6 = 1/3.

   c   P(C is ranked first and D is ranked last) = 1/6.

6.27   **a**   The four simple events are (brand M, 2 heads), (brand M, 4 heads), (brand Q, 2 heads), (brand Q, 4 heads).

   b   P(brand Q with 2 heads) = 32% = 0.32.

   c   P(brand M) = 25% + 16% = 41% = 0.41.

Exercises 6.29 – 6.39

6.29   **a** $P(E \mid F) = \dfrac{P(E \cap F)}{P(F)} = \dfrac{0.54}{0.6} = 0.90.$

   b $P(F \mid E) = \dfrac{P(E \cap F)}{P(E)} = \dfrac{0.54}{0.7} = 0.7714.$

6.31   The statement in the article implies the following two conditions:
(1) $P(D \mid Y^c) > P(D \mid Y)$ and

$$(2)\ \ P(Y) > P(Y^c).$$

This claim is consistent with the information given in I but not with any of the others. In II and III, condition (1) is violated; in IV, condition (2) is violated; in V and VI, both conditions are violated.

6.33 **a.** $P(F|F^P) = \dfrac{432}{562} = 0.769$

      b. $P(M|M^P) = \dfrac{390}{438} = 0.890$

      c. Ultrasound appears to be more reliable in predicting boys

6.35 **a** **i** Required conditional probability $= \dfrac{135}{182} = 0.74176.$

          ii Required conditional probability $= \dfrac{173+206}{210+231} = 0.8594.$

      b

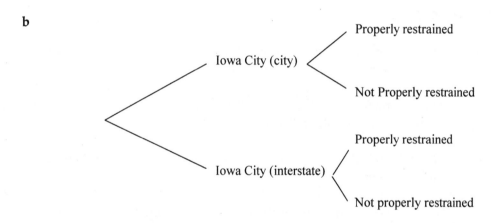

6.37 **a** $0.1 + 0.175 = 0.275$

      b $\dfrac{0.1}{0.5} = 0.2$

      c $\dfrac{0.325}{0.5} = 0.65$

      d $\dfrac{0.325}{0.725} = 0.448$

      e The probabilities in **c** and **d** are not equal because the events "wearing a seatbelt" and "being female" are not independent. It is clearly more likely an adult will regularly use a seatbelt if they are female than if they are male.

6.39 **a**   **i.** $P(S) = \dfrac{456}{600} = 0.76$ **ii.** $P(S|A) = \dfrac{215}{300} = 0.717$ **iii.** $P(S|B) = \dfrac{241}{300} = 0.803$

iv Drug B appears to have a higher survival rate

b   **i** $P(S) = \dfrac{140}{240} = 0.583$ **ii** $P(S|A) = \dfrac{120}{200} = 0.600$ **iii** $P(S|B) = \dfrac{20}{40} = 0.5$

iv Drug A appears to have the higher survival rate.

c.   **i** $P(S) = \dfrac{316}{360} = 0.878$ **ii** $P(S|A) = \dfrac{95}{100} = 0.95$ **iii** $P(S|B) = \dfrac{221}{260} = 0.85$

iv Drug A appears to have the higher survival rate

d.   Women seem to respond well to both treatments. Men seem to respond very well to treatment A but not so well to treatment B. When the data is combined, the small quantity of data collected for men on treatment B - only 20% of those men who had treatment A, becomes "lost" in the rest of the data.

Exercises 6.40 – 6.58

6.41   **a**   In the long run, 30 out of every 100 accidents reported to the police that result in a death were a single-vehicle rollover.

b   In the long run, 54 out of every 100 accidents reported to the police that result in a death were a frontal collision.

c $P(R|D) = 0.3$, $P(R) = 0.06$. As $P(R|D) \neq P(R)$, the events R and D are not independent.

d $P(F \cap D)$ is only equal to $P(F)P(D)$ if the events F and D are independent. As $P(F|D) \neq P(F)$: $0.54 \neq 0.6$, the events F and D are not independent $\Rightarrow P(F \cap D) \neq P(F)P(D)$.

e   No, F is the event that the selected accident was a frontal collision. R^c is the event that the selected accident was not a single-vehicle rollover. These are not the same. There are accidents that are one of these, or both or neither.

6.43   Let A = the event that the selected student (from all kindergartners who were tested for TB) has TB; B = the event that the selected student (selected from all kindergartners who were tested for TB) is a recent immigrant. Based on the information given, we have $P(A \mid B) = 0.0075$ and $P(A) = 0.0006$. For independence, we need $P(A \mid B) = P(A)$ which is not true here. So the two outcomes under consideration are dependent.

6.45 Let A = the event that the selected smoker who is trying to quit uses a nicotine aid and B = the event that the selected smoker who has attempted to quit begins smoking again within two weeks. From the given information we can deduce the following.

$$P(A) = 0.113; \; P(B\,|\,A^c) = 0.62; \; P(B\,|\,A) = 0.6 \,. \text{ So } P(A \cap B) = P(A)P(B\,|\,A) = 0.0678.$$

From this we get,

$$P(A^c \cap B) = P(A^c)P(B\,|\,A^c) = (1 - 0.113)(0.62) = 0.54994.$$

Therefore

$$P(B) = P(A \cap B) + P(A^c \cap B) = 0.61774.$$

Since $P(B\,|\,A)$ is not equal to $P(B)$, A and B are dependent outcomes.

6.47 a Assuming that whether Jeanie forgets to do one of her 'to do' list items is independent of whether or not she forgets any other of her 'to do' list items, the probability that she forgets all three errands = (0.1)(0.1)(0.1) = 0.001.

  b P(remembers at least one of her three errands) = 1 – P(she forgets all three errands) = 1 – 0.001 = 0.999.

  c P(remembers the first errand, but not the second or the third) = (0.9)(0.1)(0.1) = 0.009.

6.49 a P(student has a Visa card) = 7000/10000 = 0.7

  b P(student has both cards) = 5000/10000 = 0.5

  c P(student has both cards | student has Visa card) = 5000/7000 = 0.714286

  d P(student has a Visa card & a Master Card) = 0.5. For the two events to be independent we must have P(student has a Visa card & a Master Card) = P(has Visa card) P(has a Master Card) = (7000/10000) (6000/10000) = 0.42 which is not equal to 0.5. So the two events are dependent.

  e If only 4200 has both cards, then the condition P(student has a Visa card & a Master Card) = P(has Visa card) P(has a Master Card) holds, so the two events are independent.

6.51 a i P(F) = 0.51
      ii P(F | C) = 0.56
      iii P(F | C^c) = 0.45
      iv P(F | O) = 0.36
      v P(F | Y) = 0.72

  b P(F | C) is not equal to P(F) so F and C are dependent.

c $P(F \mid O)$ is not equal to $P(F)$ so F and O are dependent.

6.53 a P(A wins both matches and B beats C) = P(A beats B) P(A beats C) P(B beats C) = (0.7)(0.8)(0.6) = 0.336.

  b P(A wins both her matches) = P(A beats B) P(A beats C) = (0.7)(0.8) = 0.56.

  c P(A loses both her matches) = P(A loses to B) P(A loses to C) = (1 – 0.7)(1 – 0.8) = 0.06.

  d There are two ways this can happen – A beats B, B beats C, C beats A OR A beats C, B beats A, and C beats B. So the required probability = (0.7) (0.6) (1 – 0.8) + (0.8) (1 – 0.7) (1 – 0.6) = 0.084 + 0.096 = 0.18.

6.55 P(small size of brand B_1) = P(small) P(brand B_1) = (0.30)(0.40) = 0.12. The other probabilities are obtained in a similar manner. The results are shown in the following table.

		Size			
		S	**M**	**L**	
Brand	**B_1**	0.12	0.20	0.08	0.40
	B_2	0.18	0.30	0.12	0.60
		0.30	0.50	0.20	1.00

6.57 a CC, CN, NC, NN.

  b P(CC) = (50/800)(50/800) = 0.00391.

  c P(CC) = P(first answer is correct) P(second answer is correct | first answer is correct) = (50/800)(49/799) = 0.00383. This is very close to the probability computed in part **b**.

Exercises 6.59 – 6.74

6.59 a P(E) = 6/10 = 0.6

  b P(F | E) = 5/9 = 0.5556

  c P(E and F) = P(E) P(F|E) = (6/10)(5/9) = 30/90 = 1/3

6.61　**a**　P(individual has to stop at at least one light) = $P(E \cup F) = P(E) + P(F) - P(E \cap F)$
　　　　= 0.4 + 0.3 - 0.15 = 0.55.

　　b　P(individual doesn't have to stop at either traffic light) = 1 - P(must stop at at least one light)
　　　　= 1 - 0.55 = 0.45 (using the result of part **a**).

　　c　P(stop at exactly one of the two lights) = P(stop at least one of the two lights) - P(stop at both lights) = 0.55 - 0.15 = 0.40.

　　d　$P(E \text{ only}) = P(E \text{ and not } F) = P(E) - P(E \cap F) = 0.4 - 0.15 = 0.25.$

6.63　The likelihood of using a cell phone is different for of each type of vehicle; for instance if you drive a van or SUV, you are more likely to use a cell phone than if you drive a pick-up truck.

6.65　**a**　**i** 0.5　**ii** 0.5　**iii** 0.99　**iv** .01

　　b　$P(TD) = P(TD \cap D) + P(TD \cap C) = P(TD \mid D)P(D) + P(TD \mid C)P(C) =$
　　　　= (.99)(.01) + (.99)(.01) = 2(.99)(.01) = .0198

　　c　$P(C \mid TD) = \dfrac{P(C)P(TD \mid P(C)}{P(TD \mid P(C) + P(TD) \mid P(D)} = \dfrac{(.99)(.01)}{2(.99)(.01)} = 0.5$

　　　　Yes, it is consistent with the argument given in the quote.

6.67　**a**　In 4 consecutive years, there are 3 years of 365 days (1095 non leap days) and 1 year of 366 days (365 non leap days and one leap day) – a total of 1461 days, of which one, Feb 29th is a leap day. Hence a leap day occurs once in 1461 days.

　　b　If babies are induced or born by C-section, they are less likely to be born at weekends or on holidays so they are not equally likely to be born on the 1461 days.

　　c　1 in 2.1 million is the probability of picking a mother and baby randomly and finding *both* to be a leap-day mom-baby ($1/1461^2$). The probability of a leap year baby becoming a leap year mom means that the mom's birthday is already fixed, so the hospital spokesperson's probability is too small.

6.69　**a**　Assume that the results of successive tests on the same individual are independent of one another. Let F_1 = the event that the selected employee tests positive on the first test and F_2 = the event that he/she tests positive on the second test. Then, P(employee uses drugs and test positive twice) = $P(E)P(F_1 \mid E)P(F_2 \mid E)$ = (0.1)(0.90)(0.90) = 0.081.

b   P(employee tests positive twice)

= P(employee uses drugs and tests positive twice)

  + P(employee doesn't use drugs and tests positive twice)

= P(E) P(F_1|E)P(F_2|E) + P(E^c) P(F_1|E^c)P(F_2|E^c)

= (0.1)(0.90)(0.90) + (0.9)(0.05)(0.05) = 0.081 + 0.00225 = 0.08325.

c   P(uses drugs | tested positive twice) =

$$\frac{P(\text{uses drugs and tests positive twice})}{P(\text{tests positive twice})} = \frac{0.081}{0.08325} = 0.97297.$$

d   P(tests negative on the first test OR tests positive on the first test but tests negative on the second test | does use drugs) = P(F_1^c | E) + P(F_1 and F_2^c |E)

= [1 − P(F_1 |E)] + P(F_1 |E) [1 − P(F_2 |E)]

= [1 − 0.9] + (0.9)(1 − 0.9) = 0.1 + 0.09 = 0.19

(using P(F_1 |E) = 0.90, from part **a** of Problem **6.68**).

e   The benefit of using a retest is that it reduces the rate of false positives from 0.05 to 0.02703 (obtained as 1 − 0.97297, using the result from part **c**), but the disadvantage is that the rate of false negatives has increased from P(F^c |E) = 0.10 to 0.19 (see part **d** above). Retests also involve additional expenses.

6.71   Based on the given information we can construct the following table.

	Basic Model	Deluxe Model	Total
Buy extended warranty	12%	30%	42%
Do not buy extended warranty	28%	30%	58%
Total	40%	60%	100%

So P(Basic model | extended warranty) = 12/42 = 0.285714.

6.73   Let A = the event that he takes the small car; B = the event that he takes the big car; C = the event that he is on time. Then, based on the given information we have P(A) = 0.75, P(B) = 0.25, P(C|A) = 0.9, and P(C|B) = 0.6. We need P(A|C). Using Bayes' Rule we get

$$P(A \mid C) = \frac{P(C \mid A)P(A)}{P(C \mid A)P(A) + P(C \mid B)P(B)} = \frac{(0.9)(0.75)}{(0.9)(0.75) + (0.6)(0.25)} = 0.8182.$$

Exercises 6.75 – 6.83

6.75   **a**   P(on time delivery in Los Angeles) = 425/500 = 0.85.

      b   P(late delivery in Washington, D.C) = (500 − 405)/500 = 95/500 = 0.19.

c   Assuming that whether or not one letter is delivered late is "independent" of the on-time delivery status of any other letter, we calculate P(both letters delivered on-time in New York) = (415/500)(415/500) = 0.6889. (The independence assumption may not be a valid assumption).

d   P(on-time delivery nationwide) = 5220/6000 = 0.87.

6.77   a   P(male) = (200+800+1500+1500+900+1500+300)/17000 = 6700/17000 = 0.3941.

b   P(agriculture) = 3000/17000 = 0.1765.

c   P(male AND from agriculture) = 900/17000 = 0.0529.

d   P(male AND not from agriculture) = 5800/17000 = 0.3412.

6.79   The results will vary from one simulation to another. The approximate probabilities given were obtained from a simulation done by computer with 100,000 trials.

a   0.71293

b   0.02201

6.81   The simulation results will vary from one simulation to another. The approximate probability should be around .8504.

6.83   a   The simulation results will vary from one simulation to another. The approximate probability should be around 0.6504.

b   The decrease in the probability of on time completion for Jacob made the biggest change in the probability that the project is completed on time.

Exercises 6.84 – 6.98

6.85   P(parcel went via A₁ and was late) = P(went via A₁)P(late | went via A₁) = (0.4)(0.02) = 0.008.

6.87   a $(.3)^5 = .00243$

b $(.3)^5 + (.2)^5 = .00243 + .00032 = .00275$

c   Select a random digit. If it is a 0, 1, or 2, then award A one point. If it is a 3 or 4, then award B one point. If it is 5, 6, 7, 8, or 9, then award A and B one-half point each. Repeat the selection process until a winner is determined or the championship ends in a draw (5 points each). Replicate this entire procedure a large number of times. The estimate of the probability that A wins the championship would be the ratio of the number of times A wins to the total number of replications.

d   It would take longer if no points for a draw are awarded. It is possible that a number of games would end in a draw and so more games would have to be played in order for a player to earn 5 points.

6.89   P(need to examine at least 2 disks) = 1 – P(get a blank disk in the first selection) = 1 – 15/25 = 0.4.

6.91   Let the five faculty members be A, B, C, D, E with teaching experience of 3, 6, 7, 10, and 14 years respectively. The two selected individuals will have a total of at least 15 years of teaching experience if the selected individuals are A and E, or B and D, or B and E, or C and D, or C and E, or D and E. Since there are a total of 10 equally likely choices and 6 out of these meet the specified requirement, P(two selected members will have a total of at least 15 years experience) = 6/10 = 0.6.

6.93   Total number of viewers = 2517.

a $P(\text{saw a PG movie}) = \dfrac{179+87}{2517} = 0.1057.$

b $P(\text{saw a PG or a PG-13 movie}) = \dfrac{420+323+179+114+87}{2517} = 0.4462.$

c $P(\text{did not see an R movie}) = \dfrac{2517-600-196-205-139}{2517} = 0.5471.$

6.95   **a**   P(neither is selected for testing | batch has 2 defectives) = (8/10)(7/9) = 56/90 = 0.6222.

b   P(batch has 2 defectives and neither selected for testing) = P(batch has 2 defectives) P(neither is selected for testing | batch has 2 defectives) = (0.20)(0.6222) = 0.1244.

c   P(neither component selected is defective) = P(batch has 0 defectives & neither selected component is defective) + P(batch has 1 defective & neither selected component is defective) + P(batch has 2 defectives & neither selected component is defective) = $(0.5)(1) + (0.3)(9/10)(8/9) + (0.2)(8/10)(7/9) = 0.5 + 0.24 + 0.1244 = 0.8644$.

6.97 $P(E \cap F \cap G \cap H) = P(E)P(F|E)P(G|E \cap F)P(H|E \cap F \cap G) = (20/25)(19/24)(18/23)(17/22) = 0.3830$. P(at least one bad bulb) = 1 − P(all 4 bulbs are good) = 1 − 0.383 = 0.617.

Exercises 7.1 – 7.10

7.1 **a**   discrete

      b   continuous

      c   discrete

      d   discrete

      e   continuous

7.3 **a**

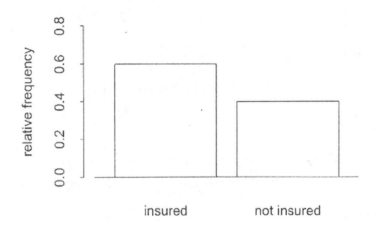

      b   P(individual does not have earthquake insurance) = 0.4

7.5 **a**

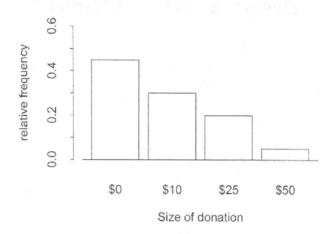

b x = 0, since its probability is the largest.

c P(x ≥ 25) = P(x = 25) + P(x = 50) = 0.20 + 0.05 = 0.25

d P(x > 0) = 1 – P(x = 0) = 1 – 0.45 = 0.55

7.7 **a** P(x ≤ 100) = 0.05 + 0.10 + 0.12 + 0.14 + 0.24 + 0.17 = 0.82

b P(x > 100) = 1 – P(x ≤ 100) = 1 – 0.82 = 0.18

c P(x ≤ 99) = 0.05 + 0.10 + 0.12 + 0.14 + 0.24 = 0.65
P(x ≤ 97) = 0.05 + 0.10 + 0.12 = 0.27

7.9 **a** Supplier 1

b Supplier 2

c Supplier 1, because the bulbs generally last longer and have less variability in their lifetimes.

d About 1000

e About 100

7.11   a

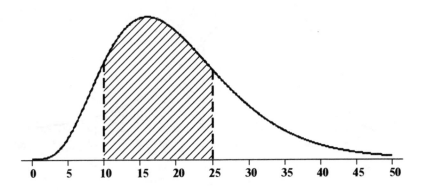

b

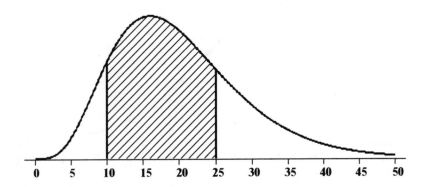

c

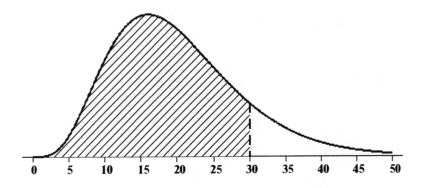

d

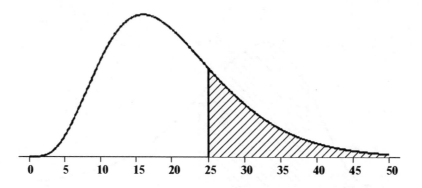

e

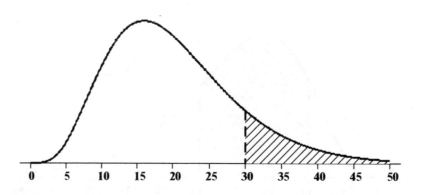

7.13 **a**   Total area = 1/2(0.40)(5) = 1

     b $P(x < 0.20) = 1/2(0.20)(5) = 0.5$
          $P(x < 0.10) = 1/2(0.10)(5/2) = 0.125$
          $P(x > 0.30) = 1/2(0.10)(5/2) = 0.125$

     c $P(0.10 \leq x \leq 0.20) = 1 - P(x < 0.10 \text{ or } x > 0.20)$
          From part **b**, $P(x < 0.10) = 0.125$ and $P(x > 0.20) = 0.5$
          Hence, $P(0.10 \leq x \leq 0.20) = 1 - (0.125 + 0.5) = 0.325$

     d $P(0.15 \leq x \leq 0.25) = 1 - [P(x \leq 0.15) + P(x \geq 0.25)]$

$$= 1 - \left[\frac{1}{2}(.15)(5)(.75) + \frac{1}{2}(.4 - .25)(5)(.75) \right]$$

$$= 1 - [0.28125 + 0.28125] = 1 - 0.5625 = 0.4375$$

7.15 **a**   The area under the density curve must equal 1. Since this area is a triangle, we
          have ½(1)h = 1. So h = 2.

b $P(x > 0.5) = (1/2)(0.5)(1) = 0.25$

c $P(x \leq 0.25) = 1 - P(x > 0.25) = 1 - [(1/2)(1 - 0.25)(1.5)] = 1 - 9/16 = 7/16$

Exercises 7.16 – 7.32

7.17   **a** $P(z < -1.28) = 0.1003$

         b $P(z > 1.28) = 1 - P(z \leq 1.28) = 1 - 0.8997 = 0.1003$

         c $P(-1 < z < 2) = P(z < 2) - P(z < -1) = 0.9772 - 0.1587 = 0.8185$

         d $P(z > 0) = 1 - P(z \leq 0) = 1 - 0.5 = 0.5$

         e $P(z > -5) = 1 - P(z \leq -5) \approx 1 - 0 = 1$

         f $P(-1.6 < z < 2.5) = P(z < 2.5) - P(z < -1.6) = 0.9938 - 0.0548 = 0.9390$

         g $P(z < 0.23) = 0.5910$

7.19   **a** $P(z < 0.1) = 0.5398$

         b $P(z < -0.1) = 0.4602$

         c $P(0.40 < z < 0.85) = P(z < 0.85) - P(z < 0.4) = 0.8023 - 0.6554 = 0.1469$

         d $P(-0.85 < z < -0.40) = P(z < -0.4) - P(z < -0.85) = 0.3446 - 0.1977 = 0.1469$

         e $P(-0.40 < z < 0.85) = P(z < 0.85) - P(z < -0.4) = 0.8023 - 0.3446 = 0.4577$

         f $P(-1.25 < z) = 1 - P(z \leq -1.25) = 1 - 0.1056 = 0.8944$

         g $P(z < -1.5 \text{ or } z > 2.5) = P(z < -1.5) + 1 - P(z \leq 2.5) = 0.0668 + 1 - 0.9938 = 0.0730$

7.21   **a** $P(z > z^*) = 0.03 \implies z^* = 1.88$

         b $P(z > z^*) = 0.01 \implies z^* = 2.33$

         c $P(z < z^*) = 0.04 \implies z^* = -1.75$

         d $P(z < z^*) = 0.10 \implies z^* = -1.28$

7.23   **a**   91st percentile = 1.34

b   77th percentile = 0.74

c   50th percentile = 0

d   9th percentile = −1.34

e   They are negatives of one another. The 100pth and 100(1−p)th percentiles will be negatives of one another, because the z curve is symmetric about 0.

7.25   **a** $P(x > 4000) = P(z > \dfrac{4000 - 3432}{482}) = P(z > 1.1784) = 0.1193$

$P(3000 \leq x \leq 4000) = P(\dfrac{3000 - 3432}{482} \leq z \leq \dfrac{4000 - 3432}{482})$

$= P(-0.8963 \leq z \leq 1.1784) = P(z \leq 1.1784) - P(z < -0.8963) = 0.8807 - 0.1851 = 0.6956$

b $P(x < 2000) + P(x > 5000) = P(z < \dfrac{2000 - 3432}{482}) + P(z > \dfrac{5000 - 3432}{482})$

$= P(z < -2.97095) + P(z > 3.25311) = 0.0015 + 0.00057 = 0.00207$

c $P(x > 7 \text{ lbs}) = P(x > 7(453.6) \text{ grams}) = P(x > 3175.2) = P(z > \dfrac{3175.2 - 3432}{482})$

$= P(z > -0.53278) = 0.70291$

d   We find x_1^* and x_2^* such that $P(x < x_1^*) = 0.0005$ and $P(x > x_2^*) = 0.0005$. The most extreme 0.1% of all birthweights would then be characterized as weights less than x_1^* or weights greater than x_2^*. $P(x < x_1^*) = P(z < z_1^*) = 0.0005$ implies that $z_1^* = -3.2905$. So $x_1^* = \mu + z_1^* \sigma = 3432 + 482(-3.2905) = 1846$ grams. $P(x > x_1^*) = P(z > z_2^*) = 0.0005$ implies that $z_2^* = 3.2905$. So $x_2^* = \mu + z_2^* \sigma = 3432 + 482(3.2905) = 5018$ grams. Hence the most extreme 0.1% of all birthweights correspond to weights less than 1846 grams or weights greater than 5018 grams.

e   If x is a random variable with a normal distribution and a is a numerical constant (not equal to 0) then $y = ax$ also has a normal distribution. Furthermore,

        mean of $y = a \times$ (mean of x)

and

        standard deviation of $y = a \times$ (standard deviation of x).

Let y be the birthweights measured in pounds. Recalling that one pound = 453.6 grams, we have $y = \left(\dfrac{1}{453.6}\right) x$, so $a = \dfrac{1}{453.6}$. The distribution of y is normal with mean equal to $3432/453.6 = 7.56614$ pounds and standard deviation equal to

$482/453.6 = 1.06261$ pounds. So $P(y > 7 \text{ lbs}) = P(z > \dfrac{7 - 7.56614}{1.06261}) = P(z > -0.53278) =$ 0.70291. As expected, this is the same answer that we obtained in part **c**.

7.27   For the second machine,

P(that a cork doesn't meet specifications) $= 1 - P(2.9 \leq x \leq 3.1)$

$= 1 - P(\dfrac{2.9 - 3.05}{0.01} \leq z \leq \dfrac{3.1 - 3.05}{0.01}) = 1 - P(-15 \leq z \leq 5) \approx 1 - 1 = 0.$

7.29 $\dfrac{c - 120}{120} = -1.28 \Rightarrow c = 120 - 1.28(20) = 94.4$

Task times of 94.4 seconds or less qualify an individual for the training.

7.31   **a** $P(x \leq 60) = P(z \leq \dfrac{60 - 60}{15}) = P(z \leq 0) = 0.5$

$P(x < 60) = P(z < 0) = 0.5$

b $P(45 < x < 90) = P(\dfrac{45 - 60}{15} < z < \dfrac{90 - 60}{15})$

$= P(-1 < z < 2) = 0.9772 - 0.1587 = 0.8185$

c $P(x \geq 105) = P(z \geq \dfrac{105 - 60}{15}) = P(z \geq 3) = 1 - P(z < 3) = 1 - 0.9987 = 0.0013$

The probability of a typist in this population having a net rate in excess of 105 is only 0.0013. Hence it would be surprising if a randomly selected typist from this population had a net rate in excess of 105.

d $P(x > 75) = P(z > \dfrac{75 - 60}{15}) = P(z > 1) = 1 - P(z \leq 1) = 1 - 0.8413 = 0.1587$

P(both exceed 75) $= (0.1587)(0.1587) = 0.0252$

e $P(z < z^{*}) = 0.20 \Rightarrow z^{*} = -0.84$

$x^{*} = \mu + z^{*}\sigma \Rightarrow x^{*} = 60 + (-0.84)(15) = 60 - 12.6 = 47.4$

So typing speeds of 47.4 words or less per minute would qualify individuals for this training.

Exercises 7.33 – 7.44

7.33   Since this plot appears to be very much like a straight line, it is reasonable to conclude that the normal distribution provides an adequate description of the steam rate distribution.

7.35

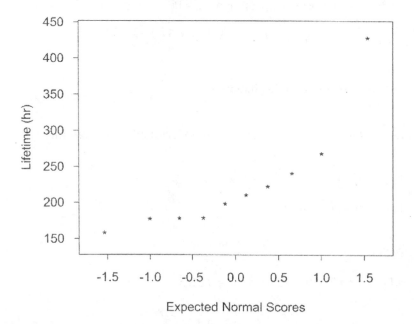

Expected Normal Scores

Since the graph exhibits a pattern substantially different from that of a straight line, one would conclude that the distribution of the variable "component lifetime" cannot be adequately modeled by a normal distribution. It is worthwhile noting that this "deviation from normality" could be due to the single outlying value of 422.6.

7.37

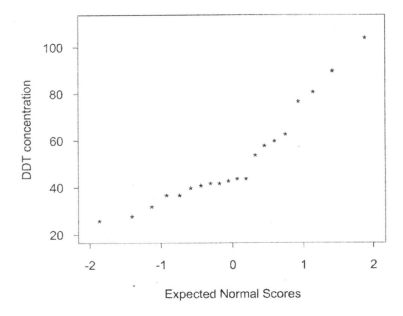

Although the graph follows a straight line pattern approximately, there is a distinct "kink" in the graph at about the value 45 on the vertical axis. Points corresponding to DDT concentration less than 45 seem to follow one straight line pattern while those to the right of 45 seem to follow a different straight line pattern. A normal distribution may not be an appropriate model for this population.

7.39 Histograms of the square root transformed data as well as the cube root transformed data are given below. It appears that the histogram of the cube root transformed data is more symmetric than the histogram of the square root transformed data. (However, keep in mind that the shapes of these histograms are somewhat dependent on the choice of class intervals.)

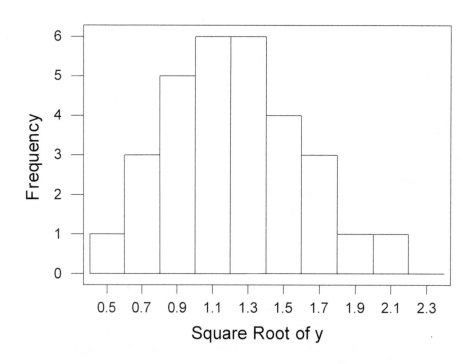

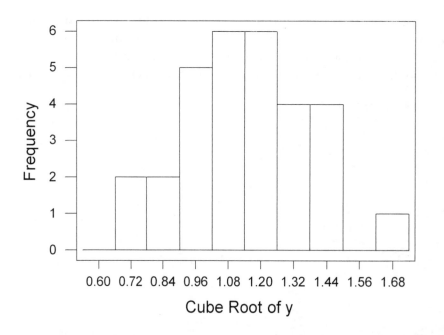

7.41   **a**   The required frequency distribution is given below.

class	frequency	relative frequency
0-<100	22	0.22
100-<200	32	0.32
200-<300	26	0.26
300-<400	11	0.11
400-<500	4	0.04
500-<600	3	0.03
600-<700	1	0.01
700-<800	0	0
800-<900	1	0.01

b

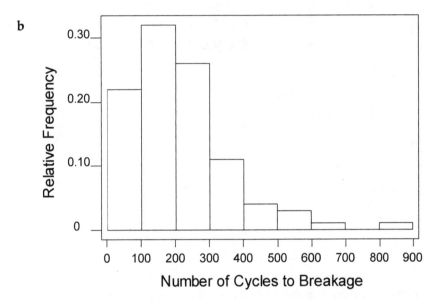

This histogram is positively skewed.

c   Data were square root transformed and the corresponding relative frequency histogram is shown below. Clearly the distribution of the transformed data is more symmetric than that of the original data.

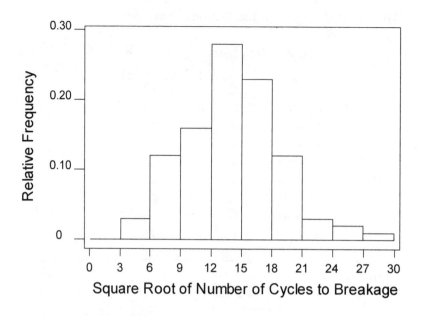

Square Root of Number of Cycles to Breakage

7.43 **a**

0	56667889999
1	000011122456679999
2	0011111368
3	11246
4	18
5	
6	8
7	
8	
9	39 HI: 448

b A density histogram of the body mass data is shown below. It is positively skewed. The scale on the horizontal axis extends upto body mass = 500 in order to accommodate the HI value of 448.

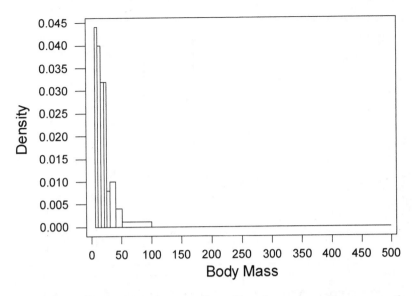

If statistical inference procedures based on normality assumptions are to be used to draw conclusions from the body mass data of this problem, then a transformation should be considered so that, on the transformed scale, the data are approximately normally distributed.

c

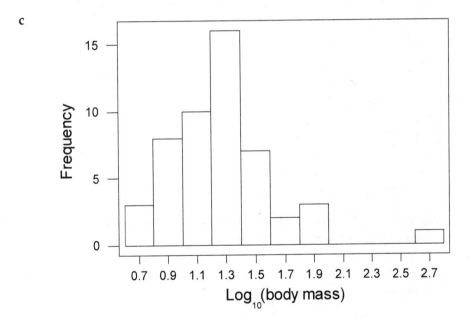

The normal curve fits the histogram of the log transformed data better than the histogram of the original data.

d

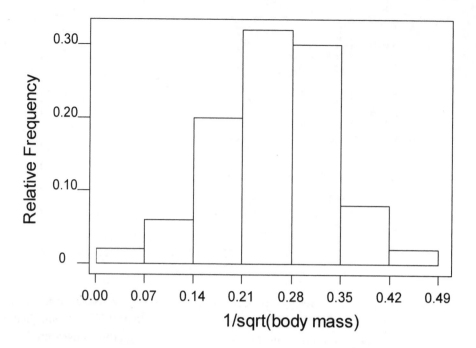

The histogram in this transformed scale does appear to have an approximate bell-shape, i.e., in the transformed scale, the data do appear to follow a normal distribution model.

Exercises 7.45 – 7.52

7.45 $P(x < 4.9) = P(z < \dfrac{4.9-5}{0.05}) = P(z < -2) = 0.0228$

$P(5.2 < x) = P(\dfrac{5.2-5}{0.05} < z) = P(4 < z) \approx 0$

7.47 **a** $P(x < 5'7") = P(x < 67") = P(z < \dfrac{67-66}{2}) = P(z < 0.5) = 0.6915$

No, the claim that 94% of all women are shorter than 5'7" is not correct. Only about 69% of all women are shorter than 5'7".

b About 69% of adult women would be excluded from employment due to the height requirement.

7.49 **a** $P(5.9 < x < 6.15) = P(\dfrac{5.9-6}{0.1} < z < \dfrac{6.15-6}{0.1}) = P(-1 < z < 1.5)$

$= 0.9332 - 0.1587 = 0.7745$

122

b $P(6.1 < x) = P(\dfrac{6.1-6}{0.1} < z) = P(1 < z) = 1 - P(z \le 1) = 1 - 0.8413 = 0.1587$

c $P(x < 5.95) = P(z < \dfrac{5.95-6}{0.1}) = P(z < -0.5) = 0.3085$

d   The largest 5% of the pH values are those pH values which exceed the 95th percentile. The 95th percentile is $6 + 1.645(0.1) = 6.1645$.

7.51 **a** $P(250 < x < 300) = P((250 - 266)/16 < z < (300 - 266)/16)$
$= P(-1 < z < 2.13) = 0.9834 - 0.1587 = 0.8247$

b $P(x < 240) = P(z < (240 - 266)/16) = P(z < -1.63) = 0.0516$

c   P(x is within 16 days of the mean duration)
$= P(250 < x < 282) = P(-1 < z < 1) = 0.8413 - 0.1587 = 0.6826$

d $P(310 \le x) = P((310 - 266)/16 < z) = P(2.75 < z)$
$= 1 - P(z \le 2.75) = 1 - 0.9970 = 0.0030$

The chances of a pregnancy having a duration of at least 310 days is 0.003. This is a small value, so there is a bit of skepticism concerning the claim of this lady.

e   If the duration is 261 days or less, then the date of birth will be $261 + 14 = 275$ days or less after coverage began. Hence the insurance company will not pay the benefits.

$P(x < 261) = P(z < (261 - 266)/16) = P(z < -0.31) = 0.3783 \approx 38\%$

When date of conception is 14 days after coverage began, about 38% of the time the insurance company will refuse to pay the benefits because of the 275 day requirement.

Peck, Olsen & Devore Chapter 7

Exercises 7.1 – 7.7

7.1 **a** discrete

  b continuous

  c discrete

  d discrete

  e continuous

7.3 The possible y values are the set of all positive integers.
Some possible outcomes are LS, RRS, S, LRRRLLRLLS, and LLLLS, with corresponding y
values equal to 2, 3, 1, 10, and 5, respectively.

7.5 Possible values of y (in feet) are the real numbers in the interval $0 \le y \le 100$. The
variable y is a continuous variable.

7.7 **a** Possible values for x are 3, 4, 5, 6, 7.

  b If y = first number – second number, then possible values of y are –3, – 2, – 1, 1, 2,
and 3.

  c Possible values of z are 0, 1, 2.

  d Possible values of w are 0, 1.

Exercises 7.8 – 7.19

7.9 **a** $p(4) = 1 - p(0) - p(1) - p(2) - p(3) = 1 - 0.65 - 0.20 - 0.10 - 0.04 = 0.01$.

  b In the long run, the proportion of cartons that have exactly one broken egg will
equal 0.20.

  c $P(y \le 2) = 0.65 + 0.20 + 0.10 = 0.95$. This means that, in the long run, the
proportion of cartons that have two or fewer broken eggs will equal 0.95.

  d $P(y < 2) = 0.65 + 0.20 = 0.85$. This probability is less than the probability in part **c**
because the event y = 2 is now not included.

e   P(exactly 10 unbroken eggs) = P(exactly 2 broken eggs) = P(y = 2) = 0.10.

f   P(at least 10 unbroken eggs) = P(0 or 1 or 2 broken eggs) = P(y ≤ 2) = 0.95 (from part **c**).

7.11   **a**   P(airline can accommodate everyone who shows up)
= P(x ≤ 100) = 0.05 + 0.10 + 0.12 + 0.14 + 0.24 + 0.17 = 0.82.

    b   P(not all passengers can be accommodated) = P(x > 100) = 1 – P(x ≤ 100) = 1 – 0.82 = 0.18.

    c   P(number 1 standby will be able to take the flight) = P(x ≤ 99)
= 0.05 + 0.10 + 0.12 + 0.14 + 0.24 = 0.65.
P(number 3 standby will be able to take the flight) = P(x ≤ 97) = 0.05 + 0.10 + 0.12
= 0.27.

7.13   Results will vary. One particular set of 50 simulations gave the following results.

Value of x	Frequency	Relative Frequency
0	13	13/50 = 0.26
1	32	32/50 = 0.64
2	5	5/50 = 0.1

These relative frequencies are quite close to the theoretical probabilities given in part **b** of Problem **7.12**.

7.15   **a**   The table below lists all possible outcomes and the corresponding x values and probabilities.

Outcome	x	Probability
FFFF	0	(0.8)(0.8)(0.8)(0.8) = 0.4096
FFFS	1	(0.8)(0.8)(0.8)(0.2) = 0.1024
FFSF	1	(0.8)(0.8)(0.2)(0.8) = 0.1024
FSFF	1	(0.8)(0.2)(0.8)(0.8) = 0.1024
SFFF	1	(0.2)(0.8)(0.8)(0.8) = 0.1024
FFSS	2	(0.8)(0.8)(0.2)(0.2) = 0.0256
FSFS	2	(0.8)(0.2)(0.8)(0.2) = 0.0256
FSSF	2	(0.8)(0.2)(0.2)(0.8) = 0.0256
SFFS	2	(0.2)(0.8)(0.8)(0.2) = 0.0256
SFSF	2	(0.2)(0.8)(0.2)(0.8) = 0.0256
SSFF	2	(0.2)(0.2)(0.8)(0.8) = 0.0256
FSSS	3	(0.8)(0.2)(0.2)(0.2) = 0.0064
SFSS	3	(0.2)(0.8)(0.2)(0.2) = 0.0064
SSFS	3	(0.2)(0.2)(0.8)(0.2) = 0.0064
SSSF	3	(0.2)(0.2)(0.2)(0.8) = 0.0064
SSSS	4	(0.2)(0.2)(0.2)(0.2) = 0.0016

From this we deduce the following probability distribution of x.

Value of x	Probability
0	0.4096
1	0.4096
2	0.1536
3	0.0256
4	0.0016

b   Both 0 and 1 are most likely values since each has probability 0.4096 of occurring.

c   P(at least 2 of the 4 have earthquake insurance) = P(x ≥ 2) = 0.1536 + 0.0256 + 0.0016 = 0.1808

7.17 a   The smallest value of y is 1. The outcome corresponding to this is S.
            The second smallest value of y is 2. The outcome that corresponds to this is FS.

b   The set of all possible y values is the set of all positive integers.

c   P(y = 1) = P(S) = 0.7
      P(y = 2) = P(FS) = (0.3)(0.7) = 0.21
      P(y = 3) = P(FFS) = (0.3)(0.3)(0.7) = $(0.3)^2 (0.7)$ = 0.063
      P(y = 4) = P(FFFS) = (0.3)(0.3)(0.3)(0.7) = $(0.3)^3 (0.7)$ = 0.0189

$$P(y = 5) = P(FFFFS) = (0.3)(0.3)(0.3)(0.3)(0.7) = (0.3)^4 (0.7) = 0.00567$$

We do see a pattern here. In fact,

$p(y) = P(y-1 \text{ failures followed by a success}) = (0.3)^{y-1} (0.7)$ for $y = 1, 2, 3, ...$

7.19   The following table lists all possible outcomes and the corresponding values of y.

Magazine 1 arrives on	Magazine 2 arrives on	Probability	Value of y
Wednesday	Wednesday	$(0.4)(0.4) = 0.16$	0
Wednesday	Thursday	$(0.4)(0.3) = 0.12$	1
Wednesday	Friday	$(0.4)(0.2) = 0.08$	2
Wednesday	Saturday	$(0.4)(0.1) = 0.04$	3
Thursday	Wednesday	$(0.3)(0.4) = 0.12$	1
Thursday	Thursday	$(0.3)(0.3) = 0.09$	1
Thursday	Friday	$(0.3)(0.2) = 0.06$	2
Thursday	Saturday	$(0.3)(0.1) = 0.03$	3
Friday	Wednesday	$(0.2)(0.4) = 0.08$	2
Friday	Thursday	$(0.2)(0.3) = 0.06$	2
Friday	Friday	$(0.2)(0.2) = 0.04$	2
Friday	Saturday	$(0.2)(0.1) = 0.02$	3
Saturday	Wednesday	$(0.1)(0.4) = 0.04$	3
Saturday	Thursday	$(0.1)(0.3) = 0.03$	3
Saturday	Friday	$(0.1)(0.2) = 0.02$	3
Saturday	Saturday	$(0.1)(0.1) = 0.01$	3

From the above table we deduce the following probability distribution for y.

Value of y	Probability
0	0.16
1	0.33
2	0.32
3	0.19

Exercises 7.20 – 7.26

7.21   **a**   P(at most 5 minutes elapses before dismissal) $= (1/10)(5 - 0) = 0.5$

b $P(3 \leq x \leq 5) = (1/10)(5 - 3) = 0.2$

7.23   **a**   The density curve for x is shown below.

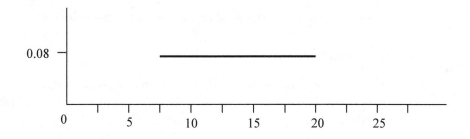

b The area under the density curve between x = 7.5 and x = 20.0 must be 1. This implies that the height of the density curve = 1/(20 – 7.5) = 1/12.5 = 0.08.

c P(x is at most 12) = (0.08)(12 – 7.5) = (0.08)(4.5) = 0.36.

d P(x is between 10 and 15) = (0.08)(15 – 10) = 0.4.
P(x is between 12 and 17) = (0.08)(17 – 12) = 0.4.
The two probabilities are equal because the distribution is uniform and therefore the probabilities depend only on the length of the interval for which the probability is being sought. This ensures they have the same area.

7.25 **a** The height of the density curve must be 1/20 = 0.05. So
P(x is less than 10 minutes) = (0.05)(10 – 0) = 0.5.
P(x is more than 15 minutes) = (0.05)(20 – 15) = 0.25.

b P(x is between 7 and 12 minutes) = (0.05)(12 – 7) = 0.25.

c P(x < c) = (0.05)(c) = 0.9 when c = 0.9/0.05 = 18. So the value of c is 18.

Exercises 7.27 – 7.44

7.27 **a** $P(x \le 0.5) = 0.5\left(\dfrac{0.5+1}{2}\right) = 0.5(0.75) = 0.375$

$P(0.25 \le x \le 0.5) = 0.25\left(\dfrac{0.75+1}{2}\right) = 0.25(0.875) = 0.21875$

$P(x \ge 0.75) = 0.25\left(\dfrac{1.25+1.5}{2}\right) = 0.25(1.375) = 0.34375$

b The mean of x is $\dfrac{7}{12} = 0.583333$ and the standard deviation of x =

$\sqrt{\dfrac{11}{144}} = 0.276385$. So,

P(x is more than one standard deviation from the mean value)
= P(x is less than (0.583333 − 0.276385)) + P(x is greater than (0.583333 + 0.276385)) = P(x is less than 0.306948) + P(x is greater than 0.859718)

$= (0.306948)\left(\dfrac{0.5 + 0.806948}{2}\right) + (1 - 0.859718)\left(\dfrac{1.359718 + 1.5}{2}\right) = 0.4012.$

7.29 **a** $\mu_y = (0)(0.65) + (1)(0.20) + (2)(0.10) + (3)(0.04) + (4)(0.01) = 0.56.$ In the long run, the average number of broken eggs per carton will equal 0.56.

b P(number of broken eggs is less than μ_y) = P(y is less than 0.56) = P(y = 0) = 0.65. So, in the long run, about 65% of the cartons will have fewer than μ_y broken eggs. This is not surprising because the distribution of y is skewed to the right and so the mean of y is greater than the median of y. We would expect the required proportion to be more than 0.5 because of the skewness.

c The indicated calculation would be correct if the values of y were all equally likely, but this is not the case. The values 0 and 1 occur more often than 2 or 3 or 4. Hence we need a "weighted average" rather than a simple average.

7.31 $\sigma_x^2 = (-1.2)^2(0.54) + (-0.2)^2(0.16) + (0.8)^2(0.06) + (1.8)^2(0.04) + (2.8)^2(0.20) = 2.52.$ Hence, $\sigma_x = \sqrt{2.52} = 1.5875.$

7.33 **a** $\mu_x = (1)(0.05) + (2)(0.10) + (3)(0.12) + (4)(0.30) + (5)(0.30) + (6)(0.11) + (7)(0.01) + (8)(0.01) = 4.12.$

b Using the definition of variance in the textbook, we get $\sigma_x^2 = 1.94560$. So $\sigma_x = \sqrt{1.94560} = 1.3948$. The average squared distance of a value of x from its mean is 1.94560. The average distance of a value of x from its mean is approximately 1.3948.

c P(x is within 1 standard deviation of the mean)
= P(x is between (4.12 − 1.3948) and (4.12 + 1.3948))
= P(x is 3, 4 or 5) = 0.12 + 0.30 + 0.30 = 0.72.

d P(x is more than 2 standard deviations from the mean) = P(x is 1, 7, or 8)
= 0.05 + 0.01 + 0.01 = 0.07.

7.35 μ_x = 10,550. The author expects to get $ 10,550 under the royalty plan whereas the flat payment is for $10,000. It would appear that the author should choose the royalty plan if he/she were quite confident about his/her assessment of the probability distribution of x. On the other hand, P(x > 10,000) = 0.25 and P(x < 10000) = 0.35, so it is more likely that the royalty plan would yield an amount less than 10,000 dollars than an amount greater than 10,000 dollars, so if the author isn't sure about his/her assessment of the probability distribution of x, then he/she might prefer the flat payment plan.

7.37   a   y is a discrete random variable because there are only 6 possible values. Successive possible values have gaps between them (this is always the case when the variable takes on only a finite number of possible values).

  b   P(paid more than $1.20 per gallon) = 0.10 + 0.16 + 0.08 + 0.06 = 0.40.
       P(paid less than $1.40 per gallon) = 0.36 + 0.24 + 0.10 + 0.16 = 0.86.

  c   The mean value of y is 126.808 (cents per gallon) and the standard deviation is 13.3162 (cents per gallon). In the long run, the average value of y will be 126.808 cents/gallon and the deviation of y on any given day, from the mean value of y, will be about 13.3162 cents/gallon.

7.39   The variable y is related to the variable x of Problem **7.38** by the relation y = 100 – 5x. Hence the mean of y is 100 – 5(Mean of x) = 100 – (5)(2.3) = 88.5. The variance of y = (25)(variance of x) = (25)(0.81) = 20.25.

7.41   a   Because if y > 0, $\Rightarrow$ x_2 > x_1 $\Rightarrow$ the diameter of the peg > the diameter of the hole and the peg wouldn't fit in the hole!

  b   0.253 – 0.25 = 0.003

  c $\sqrt{0.002^2 + 0.006^2} = 0.006$

  d   Yes, it is reasonable to assume they are independent, they are made by different tools and are randomly selected.

  e   With a standard deviation large than a mean, it seems fairly likely to obtain a negative value of y so it would seem a relatively common occurrence to find a peg that was too big to fit in the pre-drilled hole.

7.43   a   Mean of x = 2.8; standard deviation of x = 1.289.

  b   Mean of y = 0.7; standard deviation of y = 0.781.

c   Let w_1 = total amount of money collected from cars. Then $w_1 = 3x$.
Mean of $w_1 = 3$ (Mean of x) = (3)(2.8) = 8.4 dollars.
Variance of $w_1 = 9$ (Variance of x) = (9)(1.66) = 14.94.

d   Let w_2 = total amount of money collected from buses. Then $w_2 = 10y$.
Mean of $w_2 = 10$ (Mean of y) = (10)(0.7) = 7 dollars.
Variance of $w_2 = 100$ (Variance of y) = (100)(0.61) = 61.

e   Let z = total number of vehicles on the ferry. Then z = x + y.
Mean of z = Mean of x + Mean of y = 2.8 + 0.7 = 3.5.
Variance of z = Variance of x + Variance of y = 1.66 + 0.61 = 2.27.

f $w = w_1 + w_2$, so Mean of w = Mean of w_1 + Mean of w_2 = 8.4 + 7 = 15.4 dollars.
Variance of w = Variance of w_1 + Variance of w_2 = 14.94 + 61 = 75.94.

Exercises 7.45 – 63

7.45 **a**   There are exactly 6 such outcomes. They are SFFFFF, FSFFFF, FFSFFF, FFFSFF, FFFFSF, FFFFFS.

      b   In a binomial experiment consisting of 20 trials, the number of outcomes with exactly 10 S's is equal to $\binom{20}{10} = 184756$. The number of outcomes with exactly 15 S's is equal to $\binom{20}{15} = 15504$. The number of outcomes with exactly 5 S's is also equal to 15504 because $\binom{20}{15} = \binom{20}{5}$.

7.47 **a** $p(4) = \binom{6}{4}\pi^4 (1-\pi)^{6-4} = (15)(0.8)^4(0.2)^2 = 0.24576$. This means, in the long run, in samples of 6 passengers selected from passengers flying a long route, the proportion of the time exactly 4 out of the 6 will sleep or rest will be close to 0.24576.

      b $p(6) = (0.8)^6 = 0.262144$.

      c $p(x \geq 4) = p(x = 4) + p(x = 5) + p(x = 6) = 0.245760 + 0.393216 + 0.262144 = 0.90112$.

7.49 **a** $p(2) = \binom{5}{2}\pi^2 (1-\pi)^{5-2} = (10)(0.25)^2(0.75)^3 = 0.26367$.

      b $P(x \leq 1) = p(0) + p(1) = 0.2373046875 + 0.3955078125 = 0.6328125$.

c $P(2 \leq x) = 1 - P(x \leq 1) = 1 - 0.6328125 = 0.3671875.$

d $P(x \neq 2) = 1 - P(x = 2) = 1 - p(2) = 1 - 0.26367 = 0.73633.$

7.51   a $P(X = 10) = (0.85)^{10} = 0.1969$

b $P(X \leq 8) = 0.4557$

c $p = 0.15, n = 500$ mean $= 75$, st. dev. $= 7.984$

d   25 is more than 3 standard deviations from the mean value of x, so yes, this is a surprising result.

7.53   Suppose the graphologist is just guessing, i.e., deciding which handwriting is which by simply a coin toss. Then there is a 50% chance of guessing correctly in a single test. The probability of getting 6 or more correct in 10 trials, simply by guessing $= p(6) + p(7) + p(8) + p(9) + p(10)$, where p(x) is the probability that a binomial random variable with n = 10 and $\pi = 0.5$ will take the value x. Using Appendix Table X, we find this probability to be $= 0.205 + 0.117 + 0.044 + 0.010 + 0.001 = 0.377$. Therefore, correctly guessing 6 or more out of 10 is not all that rare even without any special abilities. So, the evidence given here is certainly not convincing enough to conclude that the graphologist has any special abilities.

7.55   a   P(at most 5 fail inspection) $= p(0) + p(1) + p(2) + p(3) + p(4) + p(5)$, where p(x) is the probability that a binomial random variable with n = 15 and $\pi = 0.3$ will take the value x. Using Appendix Table X, we get P(at most 5 fail inspection) $= 0.005 + 0.030 + 0.092 + 0.170 + 0.218 + 0.207 = 0.722$.

b   P(between 5 and 10 (inclusive) fail inspection) $= p(5) + p(6) + p(7) + p(8) + p(9) + p(10) = 0.207 + 0.147 + 0.081 + 0.035 + 0.011 + 0.003 = 0.484$.

c   Here, let x = number of cars that pass inspection. Then x is a binomial random variable with n = 25 and $\pi = 1 - 0.3 = 0.7$. Hence the expected value of x is $(25)(0.7) = 17.5$ and the standard deviation is $\sqrt{25(0.7)(1 - 0.7)} = 2.2913$.

7.57   Here n/N = 1000/10000 = 0.1 which is greater than 0.05. So a binomial distribution is not a good approximation for x = number of invalid signatures in a sample of size 1000 since the sampling is done without replacement.

7.59   a   P(program is implemented | $\pi = 0.8$) $= P(x \leq 15)$ where x is a binomial random variable with n = 25 and $\pi = 0.8$. Using Appendix Table X, we calculate this probability to be 0.17.

b   P(program not implemented | $\pi = 0.7$) = P(x > 15) where x is a binomial random variable with n = 25 and $\pi = 0.7$. Using Appendix Table X, we calculate this probability to be 0.811.

P(program not implemented | $\pi = 0.6$) = P(x > 15) where x is a binomial random variable with n = 25 and $\pi = 0.6$. Using Appendix Table X, we calculate this probability to be 0.425.

c   Suppose the value 15 is changed to 14 in the decision criterion.

Then P(program is implemented | $\pi = 0.8$) = P(x $\leq$ 14) where x is a binomial random variable with n = 25 and $\pi = 0.8$. Using Appendix Table X, we calculate this probability to be 0.0.006.

P(program not implemented | $\pi = 0.7$) = P(x > 14) where x is a binomial random variable with n = 25 and $\pi = 0.7$. Using Appendix Table X, we calculate this probability to be 0.902.

P(program not implemented | $\pi = 0.6$) = P(x > 14) where x is a binomial random variable with n = 25 and $\pi = 0.6$. Using Appendix Table X, we calculate this probability to be 0.586.

The modified decision criterion leads to a lower probability of implementing the program when it need not be implemented and a higher probability of not implementing the program when it should be implemented.

7.61   **a**   Geometric distribution. We are not counting the number of successes in a fixed number of trials; instead, we are counting the number of trials needed to achieve a single success.

         b   P(exactly two tosses) = (0.9)(0.1) = 0.09.

         c   P(more than three tosses will be required) = P(first three attempts are failures) = $(0.9)^3 = 0.729$.

7.63   **a**   Geometric distribution
         b   P(X = 3) = 0.1084
         c   P(X < 4) = P(X $\leq$ 3) = 0.3859
         d   P(X > 3) = 1 – P(X $\leq$ 3) = 1 - 0.3859 = 0.6141

Exercises 7.64 – 7.80

7.65   **a**   P(z < –1.28) = 0.1003

b $P(z > 1.28) = 1 - P(z \le 1.28) = 1 - 0.8997 = 0.1003$

c $P(-1 < z < 2) = P(z < 2) - P(z < -1) = 0.9772 - 0.1587 = 0.8185$

d $P(z > 0) = 1 - P(z \le 0) = 1 - 0.5 = 0.5$

e $P(z > -5) = 1 - P(z \le -5) \approx 1 - 0 = 1$

f $P(-1.6 < z < 2.5) = P(z < 2.5) - P(z < -1.6) = 0.9938 - 0.0548 = 0.9390$

g $P(z < 0.23) = 0.5910$

7.67 **a** $P(z < 0.1) = 0.5398$

b $P(z < -0.1) = 0.4602$

c $P(0.40 < z < 0.85) = P(z < 0.85) - P(z < 0.4) = 0.8023 - 0.6554 = 0.1469$

d $P(-0.85 < z < -0.40) = P(z < -0.4) - P(z < -0.85) = 0.3446 - 0.1977 = 0.1469$

e $P(-0.40 < z < 0.85) = P(z < 0.85) - P(z < -0.4) = 0.8023 - 0.3446 = 0.4577$

f $P(-1.25 < z) = 1 - P(z \le -1.25) = 1 - 0.1056 = 0.8944$

g $P(z < -1.5 \text{ or } z > 2.5) = P(z < -1.5) + 1 - P(z \le 2.5) = 0.0668 + 1 - 0.9938 = 0.0730$

7.69 **a** $P(z > z^*) = 0.03 \implies z^* = 1.88$

b $P(z > z^*) = 0.01 \implies z^* = 2.33$

c $P(z < z^*) = 0.04 \implies z^* = -1.75$

d $P(z < z^*) = 0.10 \implies z^* = -1.28$

7.71 **a** 91st percentile = 1.34
b 77th percentile = 0.74
c 50th percentile = 0
d 9th percentile = −1.34
e They are negatives of one another. The 100pth and 100(1−p)th percentiles will be negatives of one another, because the z curve is symmetric about 0.

7.73 **a** $P(x > 4000) = P(z > \dfrac{4000 - 3432}{482}) = P(z > 1.1784) = 0.1193$

$P(3000 \le x \le 4000) = P(\dfrac{3000 - 3432}{482} \le z \le \dfrac{4000 - 3432}{482})$

$= P(-0.8963 \le z \le 1.1784) = P(z \le 1.1784) - P(z < -0.8963) = 0.8807 - 0.1851 = 0.6956$

b $P(x < 2000) + P(x > 5000) = P(z < \dfrac{2000 - 3432}{482}) + P(z > \dfrac{5000 - 3432}{482})$

$= P(z < -2.97095) + P(z > 3.25311) = 0.0015 + 0.00057 = 0.00207$

c $P(x > 7 \text{ lbs}) = P(x > 7(453.6) \text{ grams}) = P(x > 3175.2) = P(z > \dfrac{3175.2 - 3432}{482})$

$= P(z > -0.53278) = 0.70291$

d We find x_1^* and x_2^* such that $P(x < x_1^*) = 0.0005$ and $P(x > x_2^*) = 0.0005$. The most extreme 0.1% of all birthweights would then be characterized as weights less than x_1^* or weights greater than x_2^*. $P(x < x_1^*) = P(z < z_1^*) = 0.0005$ implies that $z_1^* = -3.2905$. So $x_1^* = \mu + z_1^*\sigma = 3432 + 482(-3.2905) = 1846$ grams. $P(x > x_1^*) = P(z > z_2^*) = 0.0005$ implies that $z_2^* = 3.2905$. So $x_2^* = \mu + z_2^*\sigma = 3432 + 482(3.2905) = 5018$ grams. Hence the most extreme 0.1% of all birthweights correspond to weights less than 1846 grams or weights greater than 5018 grams.

e If x is a random variable with a normal distribution and a is a numerical constant (not equal to 0) then $y = ax$ also has a normal distribution. Furthermore, mean of y = $a \times$ (mean of x) and standard deviation of y = $a \times$ (standard deviation of x). Let y be the birthweights measured in pounds. Recalling that one pound = 453.6 grams, we have $y = \left(\dfrac{1}{453.6}\right) x$, so $a = \dfrac{1}{453.6}$. The distribution of y is normal with mean equal to 3432/453.6 = 7.56614 pounds and standard deviation equal to 482/453.6 = 1.06261 pounds. So $P(y > 7 \text{ lbs}) = P(z > \dfrac{7 - 7.56614}{1.06261}) = P(z > -0.53278) = 0.70291$. As expected, this is the same answer that we obtained in part **c**.

7.75 For the second machine,

P(that a cork doesn't meet specifications) = $1 - P(2.9 \le x \le 3.1)$

$$= 1 - P(\frac{2.9 - 3.05}{0.01} \le z \le \frac{3.1 - 3.05}{0.01}) = 1 - P(-15 \le z \le 5) \approx 1 - 1 = 0.$$

7.77 $\frac{c - 120}{120} = -1.28 \Rightarrow c = 120 - 1.28(20) = 94.4$

Task times of 94.4 seconds or less qualify an individual for the training.

7.79 **a** $P(x \le 60) = P(z \le \frac{60 - 60}{15}) = P(z \le 0) = 0.5$

$P(x < 60) = P(z < 0) = 0.5$

b $P(45 < x < 90) = P(\frac{45 - 60}{15} < z < \frac{90 - 60}{15})$

$= P(-1 < z < 2) = 0.9772 - 0.1587 = 0.8185$

c $P(x \ge 105) = P(z \ge \frac{105 - 60}{15}) = P(z \ge 3) = 1 - P(z < 3) = 1 - 0.9987 = 0.0013$

The probability of a typist in this population having a net rate in excess of 105 is only 0.0013. Hence it would be surprising if a randomly selected typist from this population had a net rate in excess of 105.

d $P(x > 75) = P(z > \frac{75 - 60}{15}) = P(z > 1) = 1 - P(z \le 1) = 1 - 0.8413 = 0.1587$

P(both exceed 75) = (0.1587)(0.1587) = 0.0252

e $P(z < z^*) = 0.20 \Rightarrow z^* = -0.84$

$x^* = \mu + z^*\sigma \Rightarrow x^* = 60 + (-0.84)(15) = 60 - 12.6 = 47.4$

So typing speeds of 47.4 words or less per minute would qualify individuals for this training.

Exercises 7.81 – 7.92

7.81 Since this plot appears to be very much like a straight line, it is reasonable to conclude that the normal distribution provides an adequate description of the steam rate distribution.

7.83

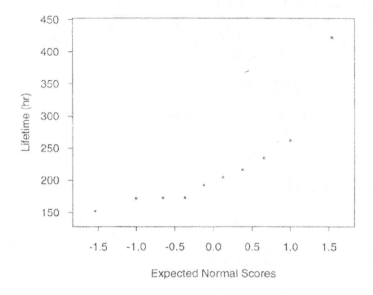

Since the graph exhibits a pattern substantially different from that of a straight line, one would conclude that the distribution of the variable "component lifetime" cannot be adequately modeled by a normal distribution. It is worthwhile noting that this "deviation from normality" could be due to the single outlying value of 422.6.

7.85

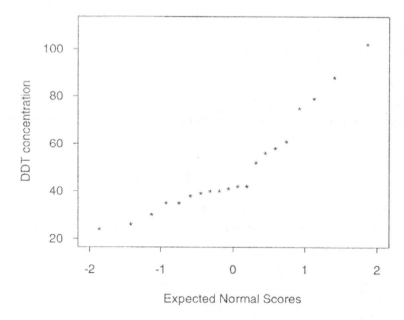

Although the graph follows a straight line pattern approximately, there is a distinct "kink" in the graph at about the value 45 on the vertical axis. Points corresponding to DDT concentration less than 45 seem to follow one straight line pattern while those to the right of 45 seem to follow a different straight line pattern. A normal distribution may not be an appropriate model for this population.

7.87 Histograms of the square root transformed data as well as the cube root transformed data are given below. It appears that the histogram of the cube root transformed data is more symmetric than the histogram of the square root transformed data. (However, keep in mind that the shapes of these histograms are somewhat dependent on the choice of class intervals.)

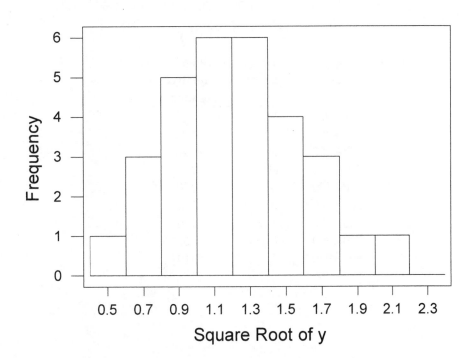

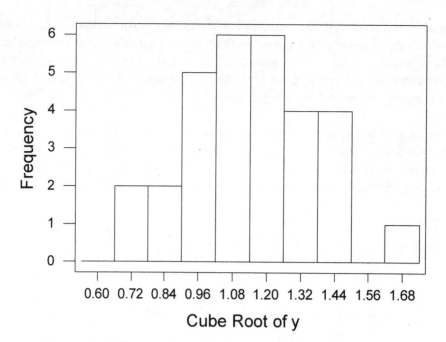

7.89 **a** The required frequency distribution is given below.

class	frequency	relative frequency
0-<100	22	0.22
100-<200	32	0.32
200-<300	26	0.26
300-<400	11	0.11
400-<500	4	0.04
500-<600	3	0.03
600-<700	1	0.01
700-<800	0	0
800-<900	1	0.01

b

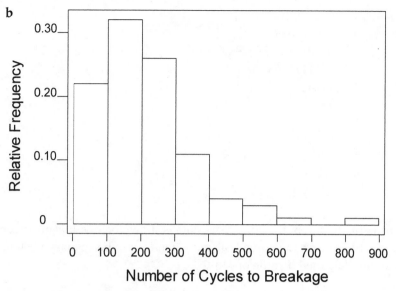

This histogram is positively skewed.

c   Data were square root transformed and the corresponding relative frequency histogram is shown below. Clearly the distribution of the transformed data is more symmetric than that of the original data.

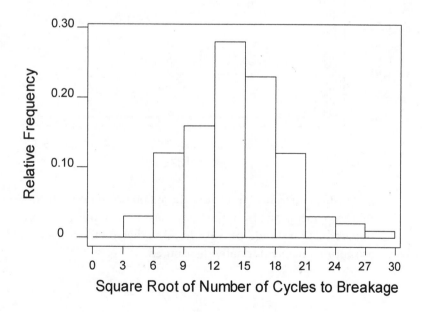

7.91   **a**

```
0    56667889999
1    000011122456679999
2    0011111368
3    11246
4    18
5
6    8
7
8
9    39    HI: 448
```

b   A density histogram of the body mass data is shown below. It is positively skewed. The scale on the horizontal axis extends upto body mass = 500 in order to accommodate the HI value of 448.

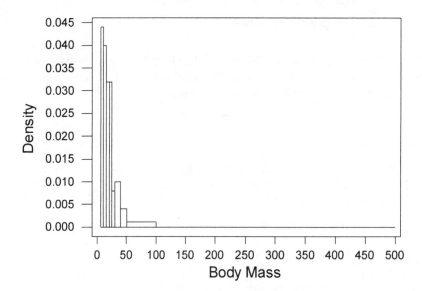

If statistical inference procedures based on normality assumptions are to be used to draw conclusions from the body mass data of this problem, then a transformation should be considered so that, on the transformed scale, the data are approximately normally distributed.

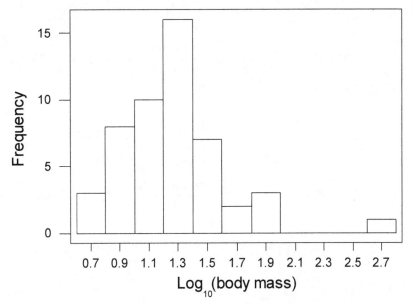

The normal curve fits the histogram of the log transformed data better than the histogram of the original data.

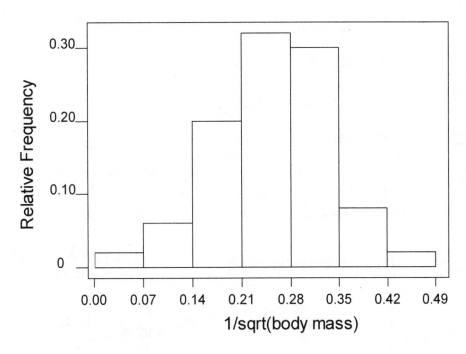

The histogram in this transformed scale does appear to have an approximate bell-shape, i.e., in the transformed scale, the data do appear to follow a normal distribution model.

Exercises 7.93 – 7.101

7.93 **a** $P(x = 100) = P\left(\dfrac{100 - 0.5 - 100}{15} \le z \le \dfrac{100 + 0.5 - 100}{15}\right) = P(-0.03333 \le z \le 0.03333)$

$= 0.02659.$

b $P(x \le 110) = P\left(z \le \dfrac{110 + 0.5 - 100}{15}\right) = P(z \le 0.7) = .75804.$

c $P(x < 110) = P(x \le 109) = P\left(z \le \dfrac{109 + 0.5 - 100}{15}\right) = P(z \le 0.6333) = 0.73674.$

d $P(75 \le x \le 125) = P\left(\dfrac{75 - 0.5 - 100}{15} \le z \le \dfrac{125 + 0.5 - 100}{15}\right) = P(-1.7 \le z \le 1.7) =$

$0.9109.$

7.95 **a** $P(650 \le x) = P\left(\dfrac{650 - 0.5 - 500}{75} \le z\right) = P(1.9933 \le z) = 0.02311.$

b $P(400 < x < 550) = P(401 \le x \le 549) = P\left(\dfrac{401 - 0.5 - 500}{75} \le z \le \dfrac{549 + 0.5 - 500}{75}\right)$

$= P(-1.3267 \le z \le 0.66) = 0.6531.$

c $P(400 \le x \le 550) = P\left(\dfrac{400 - 0.5 - 500}{75} \le z \le \dfrac{550 + 0.5 - 500}{75}\right)$

$= P(-1.34 \le z \le 0.6733) = 0.6595.$

7.97 Let x = number of mountain bikes in a sample of 100. Then x has a binomial distribution with n = 100 and $\pi = 0.7$. Its mean is equal to 70 and standard deviation is equal to $\sqrt{100(0.7)(0.3)} = 4.58258.$

a $P(\text{at most 75 mountain bikes}) = P(x \le 75) = P\left(z \le \dfrac{75 + 0.5 - 70}{4.58258}\right) = P(z \le 1.2002) =$

$0.88497.$

b $P(60 \le x \le 75) = P\left(\dfrac{60 - 0.5 - 70}{4.58258} \le z \le \dfrac{75 + 0.5 - 70}{4.58258}\right) = P(-2.29129 \le z \le 1.2002) =$

$0.8740.$

c $P(80 < x) = P(81 \le x) = P\left(\dfrac{81 - 0.5 - 70}{4.58258} \le z\right) = P(2.29129 \le z) = 0.01097.$

d   P(at most 30 are not mountain bikes) = P(at least 70 are mountain bikes) = P(70 ≤

$$x) = P\left(\frac{70 - 0.5 - 70}{4.58258} \le z \right) = P(-0.1091 \le z) = 0.54344.$$

7.99   Let x = number of voters in a random sample of size 225 that favors a 7-day waiting period. Then x is a binomial random variable with n = 225 and $\pi = 0.65$. Its mean is 146.25 and standard deviation is 7.15454.

a $$P(150 \le x) = P\left(\frac{150 - 0.5 - 146.25}{7.15454} \le z \right) = P(0.45426 \le z) = 0.32482.$$

b $$P(150 < x) = P(151 \le x) = P\left(\frac{151 - 0.5 - 146.25}{7.15454} \le z \right) = P(.59403 \le z) = 0.27625.$$

c $$P(x < 125) = P(x \le 124) = \left(z \le \frac{124 + 0.5 - 146.25}{7.15454} \right) = P(z \le -3.04003) = 0.001183.$$

7.101   **a**   Let x = number of mufflers replaced under the warranty in a random sample of 400 purchases. Then x is binomial with n = 400 and $\pi = 0.2$ Its mean is 80 and standard deviation is 8.

$$P(75 \le x \le 100)= P\left(\frac{75 - 0.5 - 80}{8} \le z \le \frac{100 + 0.5 - 80}{8} \right) = P(-0.6875 \le z \le 2.5625) = 0.74892.$$

b $$P(x \le 70) = P\left(z \le \frac{70 + 0.5 - 80}{8} \right) = P(z \le -1.1875) = 0.11752.$$

c $$P(x < 50) = P(x \le 49) = P\left(z \le \frac{49 + 0.5 - 80}{8} \right) = P(z \le -3.8125) = 0.00006878.$$ This probability is so close to zero that, if indeed the 20% figure were true, it is highly unlikely that fewer than 50 mufflers among the 400 randomly selected purchases were ever replaced. Therefore the 20% figure is highly suspect.

Exercises 7.102 – 7.124

7.103   Let x = number of customers out of the 15 who want the diet Coke. Then x is binomial with n = 15 and $\pi = 0.6$. P(each of the 15 is able to purchase the drink desired) = P(5 ≤ x ≤ 10) = 0.025 + 0.061 + 0.118 + 0.177 + 0.207 + 0.196 = 0.784 (using Appendix Table X).

7.105   **a**   Mean of x = 2.64000; standard deviation = 1.53961.

b   P(x is farther than 3 standard deviations from its mean)

= P(x is less than -1.97883 OR x is greater than 7.25883) = 0.

7.107 **a** $P(0.525 \le y \le 0.550) = P\left(\dfrac{0.525-0.5}{0.025} \le z \le \dfrac{0.575-0.5}{0.025}\right) = P(1 \le z \le 2) = 0.1359.$

b P(y exceeds its mean value by more than 2 standard deviations) = P(2 < z) = 0.02275.

c Let π = P(a randomly selected pizza has at least 0.475 lb of cheese) = P(0.475 $\le$ y) = P($-1 \le$ z) = 0.84134. Let x = number of pizzas in a random sample of size 3 that have at least 0.475 lb of cheese. Then x is a binomial random variable with n = 3 and π = 0.84134. Therefore the probability that all 3 chosen pizzas have at least 0.475 lb of cheese = P(x = 3) = $(0.84134)^3$ = 0.59555.

7.109 **a** P(y > 45) = P(z > -1.5) = 0.9332.

b Let K denote the amount exceeded by only 10% of all clients at a first meeting. Then $0.10 = P(y > K)=P\left(z > \dfrac{K-60}{10}\right)$ implies that $\dfrac{K-60}{10}$ = 1.2816. So K = 72.816 minutes.

c Let R = revenue. Then R = 10 + 50 $\left(\dfrac{y}{60}\right)$. So mean value of R = 10 + 50

$\left(\dfrac{\text{Mean of } y}{60}\right)$ = 10 + 50 $\left(\dfrac{60}{60}\right)$ = 60 dollars.

7.111 $P(x < 4.9) = P(z < \dfrac{4.9-5}{0.05}) = P(z < -2) = 0.0228.$ $P(5.2 < x) = P(\dfrac{5.2-5}{0.05} < z) = P(4 < z) \approx 0$

7.113 **a** $P(x < 5'7") = P(x < 67") = P(z < \dfrac{67-66}{2}) = P(z < 0.5) = 0.6915$

No, the claim that 94% of all women are shorter than 5'7" is not correct. Only about 69% of all women are shorter than 5'7".

b About 69% of adult women would be excluded from employment due to the height requirement.

7.115 **a** The 36 possible outcomes and the corresponding values of w are listed in the following table. The notation (i, j) means that Allison arrived at time i P.M and Teri arrived at time j P.M.

Outcome	w	Outcome	w	Outcome	w	Outcome	w
(1,1)	0	(2,4)	2	(4,1)	3	(5,4)	1

(1,2)	1	(2,5)	3	(4,2)	2	(5,5)	0
(1,3)	2	(2,6)	4	(4,3)	1	(5,6)	1
(1,4)	3	(3,1)	2	(4,4)	0	(6,1)	5
(1,5)	4	(3,2)	1	(4,5)	1	(6,2)	4
(1,6)	5	(3,3)	0	(4,6)	2	(6,3)	3
(2,1)	1	(3,4)	1	(5,1)	4	(6,4)	2
(2,2)	0	(3,5)	2	(5,2)	3	(6,5)	1
(2,3)	1	(3,6)	3	(5,3)	2	(6,6)	0

From the above table we deduce the following probability distribution of w.

w	p(w)
0	6/36
1	10/36
2	8/36
3	6/36
4	4/36
5	2/36

b The expected value of w is equal to 70/36 = 1.9444 hours.

7.117 a We use L to denote that Lygia won a game and B to denote Bob won a game. By a sequence such as BLLB we mean that Bob won the first and the fourth games and Lygia won the second and the third games. Then x = 4 occurs if either BBBB or LLLL occurs. So p(4) = $(0.6)^4 + (0.4)^4$ = 0.1552.

b The outcomes that lead to x = 5 are, LLLBL, LLBLL, LBLLL, BLLLL, BBBLB, BBLBB, BLBBB, LBBBB. Therefor p(5) = 4 $(0.6)^4(0.4)$ + 4 $(0.6)(0.4)^4$ = 0.2688.

c The maximum value of x is 7 and the minimum value of x is 4. The following table gives the distribution of x.

x	p(x)
4	$(0.6)^4 + (0.4)^4 = 0.1552$
5	$\binom{4}{3}(0.6)^4(0.4) + \binom{4}{3}(0.6)(0.4)^4 = 0.26880$
6	$\binom{5}{3}(0.6)^4(0.4)^2 + \binom{5}{3}(0.6)^2(0.4)^4 = 0.29952$
7	$\binom{6}{3}(0.6)^4(0.4)^3 + \binom{6}{3}(0.6)^3(0.4)^4 = 0.27648$

d   The expected value of x is equal to $(4)(0.1552)+(5)(0.2688) + (6)(0.29952) + (7)(0.27648) = 5.69728$.

7.119   x = number among three randomly selected customers who buy brand W. In the following, we use the sequence such as DWP to mean that the first customer buys brand D, the second buys brand W, and the third buys brand P, etc.

$P(x = 0) = P$(each of the three customers failed to buy brand W) $= (1 - 0.4)^3 = 0.216$.
$P(x = 1) = P$(exactly one of 3 customers bought brand W) $= 3(0.4)(0.6)^2 = 0.432$.
$P(x = 2) = P$(exactly two of 3 customers bought brand W) $= 3(0.4)^2(0.6) = 0.288$.
$P(x = 3) = P$(all three bought brand W) $= (0.4)^3 = 0.064$.

7.121 **a** $P(5.9 < x < 6.15) = P(\dfrac{5.9 - 6}{0.1} < z < \dfrac{6.15 - 6}{0.1}) = P(-1 < z < 1.5)$

           $= 0.9332 - 0.1587 = 0.7745$

  b $P(6.1 < x) = P(\dfrac{6.1 - 6}{0.1} < z) = P(1 < z) = 1 - P(z \le 1) = 1 - 0.8413 = 0.1587$

  c $P(x < 5.95) = P(z < \dfrac{5.95 - 6}{0.1}) = P(z < -0.5) = 0.3085$

  d   The largest 5% of the pH values are those pH values which exceed the 95th percentile. The 95th percentile is $6 + 1.645(0.1) = 6.1645$.

7.123 **a**   Let x= number among the 200 who are uninsured. Then x is binomial with n= 200 and $\pi = 0.16$. Its mean is $(200)(0.16) = 32$ and standard deviation is $\sqrt{(200)(0.16)(1 - 0.16)} = 5.18459$.

  b $P(25 \le x \le 40) = P\left(\dfrac{25 - 0.5 - 32}{5.18459} \le z \le \dfrac{40 + 0.5 - 32}{5.18459}\right) = P(-1.44659 \le z \le 1.63947)$

= 0.87544.

c $P(x > 50) = P(51 \leq x) = P\left(\dfrac{51 - 0.5 - 32}{5.18459} \leq z \right) = P(3.56827 \leq z) = 0.0001797.$ This

probability is very close to zero, so if this outcome occurs, one would be led to doubt the 16% figure.

Chapter 8

Exercises 8.1 – 8.13

8.1 A statistic is any quantity computed from the observations in a sample. A population characteristic is a quantity which describes the population from which the sample was taken.

8.3 **a** population characteristic **b** statistic
c population characteristic **d** statistic **e** statistic

8.5 We selected a random sample of size n=2 from the population of Exercise 8.4 fifty times and calculated the sample mean for each sample. The results are shown in the table below. Your own simulation may produce different results.

Trial number	First sample value	Second sample value	Sample mean	Trial number	First sample value	Second sample value	Sample mean
1	4	1	2.5	26	3	2	2.5
2	4	3	3.5	27	3	1	2.0
3	2	4	3.0	28	2	3	2.5
4	4	2	3.0	29	2	1	1.5
5	3	1	2.0	30	4	2	3.0
6	4	3	3.5	31	4	1	2.5
7	3	1	2.0	32	2	3	2.5
8	4	2	3.0	33	4	3	3.5
9	2	3	2.5	34	3	2	2.5
10	1	4	2.5	35	4	1	2.5
11	4	3	3.5	36	1	4	2.5
12	1	4	2.5	37	1	4	2.5
13	3	1	2.0	38	1	2	1.5
14	3	4	3.5	39	4	1	2.5
15	2	3	2.5	40	2	4	3.0
16	3	2	2.5	41	4	3	3.5
17	4	2	3.0	42	3	2	2.5
18	2	3	2.5	43	2	4	3.0
19	3	2	2.5	44	1	2	1.5
20	1	2	1.5	45	1	4	2.5
21	4	3	3.5	46	3	4	3.5
22	2	1	1.5	47	3	4	3.5
23	3	2	2.5	48	1	2	1.5
24	4	1	2.5	49	4	1	2.5
25	2	1	1.5	50	4	1	2.5

The relative frequency distribution of the 50 simulated values of $\bar{x}$ and the corresponding probabilities from the sampling distribution of $\bar{x}$ are as follows.

Value of $\bar{x}$	Relative frequency from simulation	Probability from sampling distribution
1.5	0.14	0.167
2	0.08	0.167
2.5	0.46	0.333
3	0.14	0.167
3.5	0.18	0.167

The relative frequencies from the simulation are approximately equal to the corresponding probabilities from the sampling distribution of $\bar{x}$. We expect the approximations to improve as the number of simulations increases. Your own simulations may lead to somewhat different results.

8.7   **a**

Sample	Value of t	Sample	Value of t
1, 5	6	5, 10	15
1, 10	11	5, 20	25
1, 20	21	10, 20	30

Value of t	6	11	15	21	25	30
Probability	1/6	1/6	1/6	1/6	1/6	1/6

b   The population mean is $\mu = (1 + 5 + 10 + 20)/4 = 36/4 = 9$

$\mu_t = (1/6)(6) + (1/6)(11) + (1/6)(15) + (1/6)(21) + (1/6)(25) + (1/6)(30) = 108/6 = 18$

μ_t is twice the value of μ. More generally, the value of μ_t will equal the value of μ times the sample size.

8.9 **Note:** Statistic #3 is sometimes called the **midrange**.

Sample	Value of Mean	Value of Median	Value of statistic #3 (midrange)
2, 3, 3*	2.67	3	2.5
2, 3, 4	3.00	3	3.0
2, 3, 4*	3.00	3	3.0
2, 3*, 4	3.00	3	3.0
2, 3*, 4*	3.00	3	3.0
2, 4, 4*	3.33	4	3.0
3, 3*, 4	3.33	3	3.5
3, 3*, 4*	3.33	3	3.5
3, 4, 4*	3.67	4	3.5
3*, 4, 4*	3.67	4	3.5

Sampling distribution of statistic #1.

Value of $\bar{x}$	2.67	3	3.33	3.67
Probability	0.1	0.4	0.3	0.2

Sampling distribution of statistic #2.

Value of median	3	4
Probability	0.7	0.3

Sampling distribution of statistic #3.

Value of midrange	2.5	3	3.5
Probability	0.1	0.5	0.4

Some of the many points to be considered when selecting an estimator of a population parameter are listed below.

i Is the estimator unbiased? i.e., is the mean of the sampling distribution of the estimator equal to the parameter being estimated?

ii Is there a high probability that the value of the estimator will be "sufficiently close" to the true parameter value?

iii Is the estimator resistant to the influence of outliers?

Your own choice of a statistic may be different from those of others.

8.11   **a** $\mu = \dfrac{8+14+16+10+11}{5} = \dfrac{59}{5} = 11.8$

b   One possible random sample consists of elements 1 and 4, whose values are 8 and 10. The value of $\bar{x}$ for this sample is $(8 + 10)/2 = 9$.

c   Twenty-four additional samples of size 2 and their means are:

Sample	Sample mean	Sample	Sample mean
14, 16	15.0	10, 16	13.0
16, 14	15.0	10, 11	10.5
11, 14	12.5	11, 14	12.5
11, 10	10.5	16, 8	12.0
14, 10	12.0	16, 8	12.0
16, 11	13.5	8, 10	9.0
14, 10	12.0	11, 10	10.5
8, 11	9.5	8, 10	9.0
16, 14	15.0	10, 8	9.0
16, 10	13.0	8, 10	9.0
10, 11	10.5	16, 14	15.0
8, 14	11.0	10, 16	13.0

d

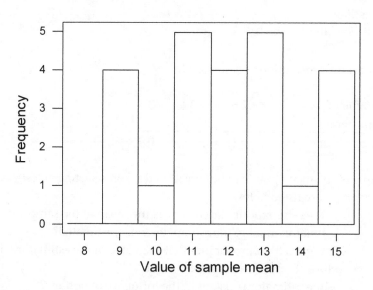

For this group of 25 samples of size two, the sample means are either close to the population mean, or at either extreme. The sample means tend to differ quite a bit from one sample to the next.

154

8.13   If samples of size ten rather than size five had been used, the histograms would be similar in that they both would be centered close to 260.25. They would differ in that the histogram based on n = 10 would have less variability than the histogram based on n = 5 (see Figure 8.4 of the text).

Exercises 8.14 – 8.26

8.15   For n = 36, 50, 100, and 400

8.17   a $\mu_{\bar{x}} = 40, \sigma_{\bar{x}} = \dfrac{5}{\sqrt{64}} = \dfrac{5}{8} = 0.625$

Since n = 64, which exceeds 30, the shape of the sampling distribution will be approximately normal.

   b $P(\mu - 0.5 < \bar{x} < \mu + 0.5) = P(39.5 < \bar{x} < 40.5) = P(\dfrac{39.5 - 40}{0.625} < z < \dfrac{40.5 - 40}{0.625})$

$= P(-0.8 < z < 0.8) = 0.7881 - 0.2119 = 0.5762$

   c $P(\bar{x} < 39.3 \text{ or } \bar{x} > 40.7) = P(z < \dfrac{39.3 - 40}{0.625} \text{ or } z > \dfrac{40.7 - 40}{0.625})$

$= 1 - P(-1.12 < z < 1.12) = 1 - [0.8686 - 0.1314] = 0.2628$

8.19   a $\mu_{\bar{x}} = 2 \text{ and } \sigma_{\bar{x}} = \dfrac{0.8}{\sqrt{9}} = 0.267$

   b   For n = 20, $\mu_{\bar{x}} = 2 \text{ and } \sigma_{\bar{x}} = \dfrac{0.8}{\sqrt{20}} = 0.179$

For n = 100, $\mu_{\bar{x}} = 2 \text{ and } \sigma_{\bar{x}} = \dfrac{0.8}{\sqrt{100}} = 0.08$

In all three cases $\mu_{\bar{x}}$ has the same value, but the standard deviation of $\bar{x}$ decreases as n increases. A sample of size 100 would be most likely to result in an $\bar{x}$ value close to μ. This is because the sampling distribution of $\bar{x}$ for n = 100 has less variability than those for n = 9 or 20.

8.21   a $\sigma_{\bar{x}} = \dfrac{5}{\sqrt{25}} = 1,$

$P(64 \le \bar{x} \le 67) = P(\dfrac{64 - 65}{1} \le z \le \dfrac{67 - 65}{1}) = P(-1 \le z \le 2) = 0.9772 - 0.1587$

$= 0.8185$

$$P(68 \le \bar{x}) = P(\frac{68-65}{1} \le z) = P(3 \le z) = 1 - P(z < 3) = 1 - 0.9987 = 0.0013$$

b   Because the sample size is large we can use the normal approximation for the sampling distribution of $\bar{x}$ by the central limit theorem. We have

$$\sigma_{\bar{x}} = \frac{5}{\sqrt{100}} = .5,$$

$$P(64 \le \bar{x} \le 67) = P((64-65)/0.5 < z < (67-65)/0.5) = P(-2 < z < 4)$$
$$= 1 - 0.0228 = 0.9772$$

$$P(68 \le \bar{x}) = P((68-65)/.5 \le z) = P(6 \le z) = 1 - P(z < 6) \approx 1 - 1 = 0$$

8.23   If the true process mean is equal to 0.5 in, then $\mu_{\bar{x}} = 0.5, \sigma_{\bar{x}} = \frac{0.02}{\sqrt{36}} = 0.00333$

P(the line will be shut down unnecessarily) = $1 - P(0.49 \le \bar{x} \le 0.51)$
$$= 1 - P(\frac{0.49-0.50}{0.00333} \le z \le \frac{0.51-0.50}{0.00333})$$
$$= 1 - P(-3 \le z \le 3) = 1 - (0.9987 - 0.0013) = 0.0026$$

8.25   The total weight of their baggage will exceed the limit, if the average weight exceeds 6000/100 = 60.

$$\mu_{\bar{x}} = 50, \quad \sigma_{\bar{x}} = \frac{20}{\sqrt{100}} = 2,$$

$$P(\text{total weight} \ge 6000) = P(\bar{x} \ge 60) = P(z \ge \frac{60-50}{2}) = P(z \ge 5) = 1 - P(z < 5) \approx 1 - 1 = 0$$

Exercises 8.27 – 8.35

8.27   **a** $\mu_p = 0.65, \sigma_p = \sqrt{0.65(0.35)/10} = 0.15083$

        b $\mu_p = 0.65, \sigma_p = \sqrt{0.65(0.35)/20} = 0.10665$

        c $\mu_p = 0.65, \sigma_p = \sqrt{0.65(0.35)/30} = 0.08708$

        d $\mu_p = 0.65, \sigma_p = \sqrt{0.65(0.35)/50} = 0.06745$

        e $\mu_p = 0.65, \sigma_p = \sqrt{0.65(0.35)/100} = 0.04770$

f $\mu_p = 0.65, \sigma_p = \sqrt{0.65(0.35)/200} = 0.03373$

8.29 **a** $\mu_p = 0.07, \sigma_p = \sqrt{\dfrac{(0.07)(0.93)}{100}} = 0.0255$

b No; $n\pi = 100(0.07) = 7$, and $n(1 - \pi) = 100(1 - 0.07) = 93$. For the sampling distribution of p to be considered approximately normal, both have to be greater or equal to 10.

c The value of the mean doesn't change as it isn't dependent on the sample size.
The standard deviation becomes : $\sigma_p = \sqrt{\dfrac{(0.07)(0.93)}{200}} = 0.01804$

d Yes; $n\pi = 200(0.07) = 14$, and $n(1 - \pi) = 200(1 - 0.07) = 186$. For the sampling distribution of p to be considered approximately normal, both have to be greater or equal to 10.

e $P(p>0.1) = P(z > \dfrac{0.1 - 0.07}{0.01804}) = P(z > 1.66) = 0.0485$

8.31 **a** $\mu_p = 0.005, \sigma_p = \sqrt{\dfrac{(0.005)(0.995)}{100}} = 0.007$

b Since $n\pi = 100(0.005) = 0.5$ is less than 10, the sampling distribution of p cannot be approximated well by a normal curve.

c The requirement is that $n\pi \geq 10$, which means that n would have to be at least 2000.

8.33 **a** For $\pi = 0.5$, $\mu_p = 0.5$ and $\sigma_p = \sqrt{\dfrac{0.5(0.5)}{225}} = 0.0333$

For $\pi = 0.6$, $\mu_p = 0.6$ and $\sigma_p = \sqrt{\dfrac{0.6(0.4)}{225}} = 0.0327$

For both cases, $n\pi \geq 10$ and $n(1 - \pi) \geq 10$. Hence, in each instance, p would have an approximately normal distribution.

b For $\pi = 0.5$, $P(p \geq 0.6) = P(z \geq \dfrac{0.6 - 0.5}{0.0333}) = P(z \geq 3) = 1 - P(z < 3)$

$= 1 - 0.9987 = 0.0013.$

For $\pi = 0.6$, $P(p \geq 0.6) = P(z \geq \dfrac{0.6 - 0.6}{0.0327}) = P(z \geq 0) = 1 - P(z < 0)$

$= 1 - 0.5000 = 0.5000$

c   When $\pi = 0.5$, the $P(p \geq 0.6)$ would decrease.
When $\pi = 0.6$, the $P(p \geq 0.6)$ would remain the same.

8.35   a $\mu_p = \pi = 0.05$, $\sigma_p = \sqrt{\dfrac{(0.05)(0.95)}{200}} = 0.01541$

$P(p > 0.02) = P(z > \dfrac{0.02 - 0.05}{0.01541}) = P(z > -1.95)$

$= 1 - P(z \leq -1.95) \approx 1 - 0.0258 = 0.9742$

b $\mu_p = \pi = 0.10$, $\sigma_p = \sqrt{\dfrac{(0.1)(0.9)}{200}} = 0.02121$

$P(p \leq 0.02) = P(z \leq \dfrac{0.02 - 0.10}{0.02121}) = P(z \leq -3.77) \approx 0$

Exercises 8.36 – 8.42

8.37   a   The sampling distribution of $\bar{x}$ will be approximately normal, with mean equal to 50 (lb) and standard deviation equal to $\dfrac{1}{\sqrt{100}} = 0.1$ (lb).

b $P(49.75 < \bar{x} < 50.25) = P(\dfrac{49.75 - 50}{0.1} < z < \dfrac{50.25 - 50}{0.1}) = P(-2.5 < z < 2.5)$

$= .9938 - .0062 = .9876$

c $P(\bar{x} < 50) = P(z < \dfrac{50 - 50}{0.1}) = P(z < 0) = 0.5$

8.39   a $\mu_{\bar{x}} = 52$ (minutes) and $\sigma_{\bar{x}} = \dfrac{2}{\sqrt{36}} = 0.33$ (minutes)

b $P(\bar{x} > 50) = P(z > \dfrac{50 - 52}{0.33}) = P(z > -6) = 1 - P(z \leq -6) \approx 1 - 0 = 1$

$$P(55 < \bar{x} > 55) = P(z > \frac{55 - 52}{0.33}) = P(z > 9) = 1 - P(z \leq 9) \approx 1 - 1 = 0$$

8.41 $\mu_p = 0.4, \sigma_p = \sqrt{\frac{(0.4)(0.6)}{100}} = 0.049$

a $P(0.5 \leq p) \approx P(\frac{0.5 - 0.4}{0.049} \leq \frac{p - 0.4}{0.049}) = P(2.04 \leq z) = 1 - P(z < 2.04)$

$= 1 - 0.9793 = 0.0207$

b The probability of obtaining a sample of 100 in which at least 60 participated in such a plan is almost 0 if $\pi = 0.40$. Hence, one would doubt that $\pi = 0.40$ and strongly suspect that π has a value greater than 0.4.

Chapter 9

Exercises 9.1 – 9.10

9.1   Statistic II would be preferred because it is unbiased and has smaller variance than the other two.

9.3   The reported number of readings above 4 pCi is 68. The point estimate of π, the proportion of all homes on the reservation whose readings exceed 4 pCi

is $p = \dfrac{68}{270} = 0.252$.

9.5   **a** $\dfrac{19.57}{10} = 1.957$

        b $s^2 = 0.15945$

        c $s = 0.3993$. No, this estimate is not unbiased. It underestimates the true value of σ.

9.7   **a** $\bar{x} = \dfrac{392.4}{19} = 20.6526$

        b   The number of cyclists in the sample whose gross efficiency is at most 20 is 4. The estimate of all such cyclists whose gross efficiency is at most 20 is the sample proportion $p = 4/19 = 0.2105$.

9.9   The point estimate of π would be $p = \dfrac{\text{number in sample registered}}{n} = \dfrac{14}{20} = 0.70$

Exercises 9.11 – 9.27

9.11   **a**   As the confidence level increases, the width of the large sample confidence interval also increases.

        b   As the sample size increases, the width of the large sample confidence interval decreases.

9.13   For the interval to be appropriate, $np \geq 10$, $n(1 - p) \geq 10$ must be satisfied.

        a $np = 50(0.3) = 15$, $n(1 - p) = 50(0.7) = 35$, yes

        b $np = 50(0.05) = 2.5$, no

c $np = 15(0.45) = 6.75$, no

d $np = 100(0.01) = 1$, no

e $np = 100(0.70) = 70$, $n(1 - p) = 100(0.3) = 30$, yes

f $np = 40(0.25) = 10$, $n(1 - p) = 40(0.75) = 30$, yes

g $np = 60(0.25) = 15$, $n(1 - p) = 60(0.75) = 45$, yes

h $np = 80(0.10) = 8$, no

9.15   a   The sample proportion p is $38/115 = 0.330$. The 95% confidence interval would be

$$0.330 \pm 1.96 \sqrt{\frac{0.330(1 - 0.330)}{115}} \Rightarrow 0.330 \pm 0.0859 \Rightarrow (0.2441, 0.4159).$$

     b   The sample proportion p is $22/115 = 0.1913$. The 90% confidence interval would

$$\text{be } 0.1913 \pm 1.645 \sqrt{\frac{0.1913(1 - 0.1913)}{115}} \Rightarrow 0.1913 \pm 0.0603 \Rightarrow (0.1310, 0.2516).$$

     c   The interval is wider in part a because (i) the confidence interval is higher in part (a) and (ii) the sample proportion is more extreme; further from 0.5 in part (a) .

9.17   a   The sample proportion who can identify their own country is 0.9. The 90%

        confidence interval would be $0.9 \pm 1.645 \sqrt{\dfrac{0.9(1 - 0.9)}{3000}} \Rightarrow 0.9 \pm 0.00901 \Rightarrow (0.891,$

        0.909).

     b   The sample should be a SRS of the respondents and independent of each other.

     c   The results would only apply to the population of respondents aged 18 to 24 in the nine different countries chosen for the study.

9.19   If the bound of error is 3.1%, the confidence interval is 36% $\pm$ 3.1%

$$p \pm z^* \sqrt{\frac{p(1 - p)}{n}} = .36 \pm .031 \Rightarrow z^* \sqrt{\frac{.36(1 - .36)}{1004}} = .031 \Rightarrow z^*(0.015) = 0.031 \Rightarrow z^* = 2.05$$

The confidence level is 96%

9.21   Based on sample data, we estimate that the percentage of all households that experienced some sort of crime during the past year is between 22% and 28%. The method used to construct this interval estimate has a 5% error rate.

9.23 The sample proportion p is 0.48. The 90% confidence interval would be

$$0.48 \pm 1.645 \sqrt{\frac{0.48(1-0.48)}{369}} \Rightarrow 0.48 \pm 0.0428 \Rightarrow (0.437, 0.523).$$

Based on this interval, we conclude with 90% confidence that the true proportion of students at The College of New Jersey who are binge drinkers is between 43.7% and 52.3%.

9.25 A 90% confidence interval is $0.65 \pm 1.645 \sqrt{\frac{0.65(1-0.65)}{150}} \Rightarrow 0.65 \pm 0.064 \Rightarrow (0.589, 0.714)$

Thus, we can be 90% confident that between 58.9% and 71.4% of Utah residents favor fluoridation. This is consistent with the statement that a clear majority of Utah residents favor fluoridation.

9.27 $n = 0.25 \left[\frac{1.96}{B} \right]^2 = 0.25 \left[\frac{1.96}{0.05} \right]^2 = 384.16;$ take n = 385.

Exercises 9.28 – 9.48

9.29
a   2.12
b   1.80
c   2.81
d   1.71
e   1.78
f   2.26

9.31 As the sample size increases, the width of the interval decreases. The interval (51.3, 52.7) has a width of 52.7 – 51.3 = 1.4 and the interval (49.4, 50.6) has a width of 50.6 – 49.4 = 1.2. Hence, the interval (49.4, 50.6) is based on the larger sample size.

9.33
a   The t critical value for a 95% confidence interval when df = 99 is 1.99. The confidence interval based on this sample data is

$$\bar{x} \pm (t\,\text{critical}) \frac{s}{\sqrt{n}} \Rightarrow 183 \pm (1.99) \left(\frac{20}{\sqrt{100}} \right) \Rightarrow (179.03, 186.97).$$

b $\bar{x} \pm (t\,\text{critical}) \frac{s}{\sqrt{n}} \Rightarrow 190 \pm (1.99) \left(\frac{23}{\sqrt{100}} \right) \Rightarrow (185.44, 194.56)$

c   The new FAA recommendations are above the upper level of both confidence levels so it appears that Frontier airlines have nothing to worry about.

9.35
a   The t critical value for a 90% confidence interval when df = 9 is 1.83. The confidence interval based on this sample data is

$$\bar{x} \pm (t\,\text{critical})\frac{s}{\sqrt{n}} \Rightarrow 54.2 \pm (1.83)\left(\frac{3.6757}{\sqrt{10}}\right) \Rightarrow 54.2 \pm 2.1271 \Rightarrow (52.073, 56.327).$$

b   If the same sampling method was used to obtain other samples of the same size and confidence intervals were calculated from these samples, 90% of them would contain the true population mean.

c   As airlines are often rated by how often their flights are late, I would recommend the published arrival time to be close to the upper bound of the confidence interval of the journey time: 10: 57 a.m.

9.37   **a** $0.5 \pm 1.96\dfrac{0.4}{\sqrt{77}} \Rightarrow 0.5 \pm 0.089 \Rightarrow (0.411, 0.589)$

b   The fact that 0 is not contained in the confidence interval does not imply that *all* students lie to their mothers. There may be students in the population and even in this sample of 77 who did not lie to their mothers. Even though the mean may not be zero, some of the individual data values may be zero. However, if the mean is nonzero, it does imply that some students tell lies to their mothers.

9.39   With n = 25, the degrees of freedom is $n - 1 = 25 - 1 = 24$.

From the t-table, the t critical value is 1.71.

The confidence interval is $2.2 \pm 1.71\left(\dfrac{1.2}{\sqrt{25}}\right) \Rightarrow 2.2 \pm 0.41 \Rightarrow (1.79, 2.61).$

9.41   The t critical value for a 90% confidence interval when df = 10 – 1 = 9 is 1.83. From the given data, n = 10, $\sum x = 219$, and $\sum x^2 = 4949.92$. From the summary statistics,

$$\bar{x} = \frac{219}{10} = 21.9$$

$$s^2 = \frac{4949.92 - \dfrac{(219)^2}{10}}{9} = \frac{4949.92 - 4796.1}{9} = \frac{153.82}{9} = 17.09$$

$$s = \sqrt{17.09} = 4.134 \,.$$

The 90% confidence interval based on this sample data is

$$\bar{x} \pm (t\ \text{critical})\frac{s}{\sqrt{n}} \Rightarrow 21.9 \pm (1.83)\frac{4.134}{\sqrt{10}} \Rightarrow 21.9 \pm 2.39 \Rightarrow (19.51,\ 24.29).$$

9.43   Summary statistics for the sample are: n = 5, $\bar{x}$ = 17, s = 9.03

The 95% confidence interval is given by

$$\bar{x} \pm (t \text{ critical}) \frac{s}{\sqrt{n}} \Rightarrow 17 \pm (2.78) \frac{9.03}{\sqrt{5}} \Rightarrow 17 \pm 11.23 \Rightarrow (5.77, \ 28.23).$$

9.45   Since the sample size is small (n = 17), it would be reasonable to use the t confidence interval only if the population distribution is normal (at least approximately). A histogram of the sample data (see figure below) suggests that the normality assumption is not reasonable for these data. In particular, the values 270 and 290 are much larger than the rest of the data and the distribution is skewed to the right. Under the circumstances the use of the t confidence interval for this problem is not reasonable.

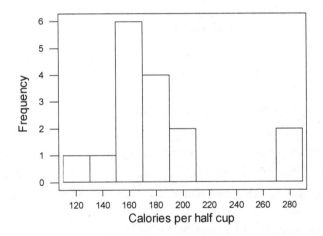

9.47 $n = \left[\dfrac{(z \, critical) \sigma}{B} \right]^2 = \left[\dfrac{(1.96)(1)}{0.1} \right]^2 = (19.6)^2 = 384.16$. Hence, n should be 385.

Exercises 9.49 – 9.64

9.49   The 95% confidence interval would be

$$0.77 \pm 1.96 \sqrt{\frac{0.77(0.23)}{800}} \Rightarrow 0.77 \pm 0.029 \Rightarrow (0.741, 0.799).$$

9.51   **a** $np = 400(0.04) = 16$ and $n(1 - p) = 400(0.96) = 384$. Since both np and n(1 – p) are greater than or equal to 5, the sample size is large enough to justify the use of the large-sample confidence interval for a population proportion.

       b   The 90% confidence interval based on this sample data is

$$p \pm (z\text{critical}) \sqrt{\frac{p(1-p)}{n}} \Rightarrow 0.04 \pm 1.645 \sqrt{\frac{0.04(0.96)}{400}} \Rightarrow 0.04 \pm 0.016 \Rightarrow (0.024, 0.056).$$

With 90% confidence, it is estimated that the actual proportion of all ABA members who are African-American is between 0.024 and 0.056.

9.53   **a**   The 95% confidence interval for μ_{ABC} is

$$15.6 \pm (1.96) \left(\frac{5}{\sqrt{50}} \right) \Rightarrow 15.6 \pm 1.39 \Rightarrow (14.21, \ 16.99).$$

       b   For μ_{CBS}: $11.9 \pm 1.39 \Rightarrow (10.51, \ 13.29)$

For μ_{FOX}: $11.7 \pm 1.39 \Rightarrow (10.31, \ 13.09)$

For μ_{NBC}: $11.0 \pm 1.39 \Rightarrow (9.61, \ 12.39)$

       c   Yes, because the plausible values for μ_{ABC} are larger than the plausible values for the other means. That is, μ_{ABC} is plausibly at least 14.21, while the other means are plausibly no greater than 13.29, 13.09, and 12.39.

9.55   Since n = 18, the degrees of freedom is $n - 1 = 18 - 1 = 17$.
From the t-table, the t critical value is 2.11.

The confidence interval is $9.7 \pm 2.11 \left(\frac{4.3}{\sqrt{18}} \right) \Rightarrow 9.7 \pm 2.14 \Rightarrow (7.56, 11.84)$.

Based on this interval we can conclude with 95% confidence that the true average rating of acceptable load is between 7.56 kg and 11.84 kg.

9.57   B = 0.1, σ = 0.8

$$n = \left[\frac{1.96\sigma}{B} \right]^2 = \left[\frac{(1.96)(0.8)}{0.1} \right]^2 = (15.68)^2 = 245.86$$

Since a partial observation cannot be taken, n should be *rounded up* to n = 246.

9.59 The 99% confidence interval for the mean commuting distance based on this sample is

$$\bar{x} \pm (\text{t critical}) \frac{s}{\sqrt{n}} \Rightarrow 10.9 \pm (2.58) \left(\frac{6.2}{\sqrt{300}} \right) \Rightarrow 10.9 \pm 0.924 \Rightarrow (9.976, 11.824).$$

9.61 $\bar{x} + 2.33 \dfrac{s}{\sqrt{n}} = 32.0 + 2.33 \left(\dfrac{22}{\sqrt{44}} \right) = 32 + 7.7277 = 39.7277$

9.63 The 95% confidence interval for the population standard deviation of commuting distance is

$$6.2 \pm 1.96 \left(\frac{6.2}{\sqrt{2(300)}} \right) \Rightarrow 6.2 \pm 0.496 \Rightarrow (5.704, 6.696).$$

Chapter 10

10.1 $\bar{x} = 50$ is not a legitimate hypothesis, because $\bar{x}$ is a statistic, not a population characteristic. Hypotheses are always expressed in terms of a population characteristic, not in terms of a sample statistic.

10.3 If we use the hypothesis H_o: $\mu = 100$ versus H_a: $\mu > 100$, we are taking the position that the welds do not meet specifications, and hence, are not acceptable unless there is substantial evidence to show that the welds are good (i.e. $\mu > 100$). If we use the hypothesis H_o: $\mu = 100$ versus H_a: $\mu < 100$, we initially believe the welds to be acceptable, and hence, they will be declared unacceptable only if there is substantial evidence to show that the welds are faulty. It seems clear that we would choose the first set-up, which places the burden of proof on the welding contractor to show that the welds meet specifications.

10.5 H_o: $\mu = 7.3$ versus H_a: $\mu > 7.3$

10.7 H_o: $\pi = 0.5$ versus H_a: $\pi > 0.5$

10.9 A majority is defined to be more than 50%. Therefore, the commissioner should test:
H_o: $\pi = 0.5$ versus H_a: $\pi > 0.5$

Exercises 10.11 – 10.21

10.11 **a**   Thinking that cancer is present when it is not means that the null hypothesis is true but it is
rejected. This is precisely the definition of type-I error.

b   A type-I error occurs in this problem when a cancer screening test result is judged to indicate presence of cancer even though the patient is actually cancer-free. The main consequence of such an error is that additional follow-up tests will be performed or treatments for cancer may be prescribed even though they are not necessary.

c   A type-II error occurs in this problem when a patient actually has cancer but the screening tests fail to indicate this. As a consequence of such an error the patient will lose the opportunity to receive timely treatment and may even ultimately die.

d   Given the same amount of data (information), any strategy to lower the risk of a type-I error will increase the risk of a type-II error.

10.13 **a**   They failed to reject the null hypothesis, because their conclusion "There is no evidence of increased risk of death due to cancer" for those living in areas with nuclear facilities is precisely what the null hypothesis states.

b   They would be making a type II error since a type I error is failing to reject the null hypothesis when the null is false.

c   Since the null hypothesis is the initially favored hypothesis and is presumed to be the case until it is determined to be false, if we fail to reject the null hypothesis, it is not proven to be true. There is just not sufficient evidence to refute its presumed truth. On the other hand, if the hypothesis test is conducted with

H_o : π is greater than the value for areas without nuclear facilities

H_a : π is less than or equal to the value for areas without nuclear facilities

and the null hypothesis is rejected, then there would be evidence based on data against the presumption of an increased cancer risk associated with living near a nuclear power plant. This is as close as one can come to "proving" the absence of an increased risk using statistical studies.

10.15 **a**   Pizza Hut's decision is consistent with the decision of rejecting H_o.

b   Rejecting H_o when it is true is called a type I error. So if they incorrectly reject H_o, they are making a type I error.

10.17 **a**   A type I error is returning to the supplier a shipment which is not of inferior quality. A type II error is accepting a shipment of inferior quality.

b   The calculator manufacturer would most likely consider a type II error more serious, since they would then end up producing defective calculators.

c   From the supplier's point of view, a type I error would be more serious, because the supplier would end up having lost the profits from the sale of the good printed circuits.

10.19 **a**   The manufacturer claims that the percentage of defective flares is 10%. Certainly one would not object if the proportion of defective flares is less than 10%. Thus, one's primary concern would be if the proportion of defective flares exceeds the value stated by the manufacturer.

b   A type I error entails concluding that the proportion of defective flares exceeds 10% when, in fact, the proportion is 10% or less. The consequence of this decision would be the filing of charges of false advertising against the manufacturer, who is not guilty of such actions. A type II error entails concluding that the proportion of defective flares is 10% when, in reality, the proportion is in excess of 10%. The consequence of this decision is to allow the manufacturer who is guilty of false advertising to continue bilking the consumer.

10.21 **a**   The manufacturer should test the hypotheses
$H_0: \pi = 0.02$ versus $H_a: \pi < 0.02$.

The null hypothesis is implicitly stating that the proportion of defective installations using robots is 0.02 or larger. That is, at least as large as for humans. In other words, they will not undertake to use robots unless it can be shown quite strongly that the defect rate *is less* for robots than for humans.

b   A type I error is changing to robots when in fact they are not superior to humans. A type II error is not changing to robots when in fact they are superior to humans.

c   Since a type I error means substantial loss to the company as well as to the human employees who would become unemployed a small α should be used. Therefore, $\alpha = 0.01$ is preferred.

Exercises 10.22 – 10.40

10.23 H_0 is rejected if P-value $\leq \alpha$. Since $\alpha = 0.05$, H_0 should be rejected for the following P-values:

a: 0.001, **b:** 0.021, and **d:** 0.047

10.25 Using Table II from the Appendix we get the following:

a   0.0808

b   0.1762

c   0.0250

d   0.0071

e   0.5675

10.27 **1.** Let π represent the proportion of U.S. adults who believe that rudeness is a worsening problem.

2. H_o: $\pi = 0.75$

3. H_a: $\pi > 0.75$

4. We will compute a P-value for this test. $\alpha = 0.05$.

5. Since $n\pi = 2013(0.75) \geq 10$, and $n(1 - \pi) = 2013(0.25) \geq 10$, the large sample z test may be used.

6. $n = 2013$, $p = 1283/2013 = 0.637357$

$$z = \frac{0.63736 - 0.75}{\sqrt{\dfrac{0.75(0.25)}{2013}}} = \frac{-.11264}{.00965} = -11.67$$

7. P-value = area under the z curve to the right of $-11.67 \approx 1$

8. Since the P-value is greater than α, H_o cannot be rejected.

9. There is not enough evidence to suggest that over ¾ of U.S. adults think that rudeness is a worsening problem.

10.29 **1.** Let π represent the proportion of religion surfers who belong to a religious community.

2. H_o: $\pi = 0.68$

3. H_a: $\pi \neq 0.68$

4. We will compute a P-value for this test. $\alpha = 0.05$.

5. Since $n\pi = 512(0.68) \geq 10$, and $n(1 - \pi) = 512(0.32) \geq 10$, the large sample z test may be used.

6. $n = 512$, $p = .84$

$$z = \frac{0.84 - .68}{\sqrt{\dfrac{0.68(0.32)}{512}}} = \frac{.16}{.0206} = 7.77$$

7. P-value = 2(area under the z curve to the right of 7.77) ≈ 2(0) = 0

8. Since the P-value is less than α, H₀ is rejected.

9. There is enough evidence to suggest that the proportion of religion surfers that belong to a religious community is different than 68%.

10.31   1.   Let π represent the proportion of the time the Belgium Euro lands with its head side up.

2. H₀: π = 0.5

3. Hₐ: π ≠ 0.5

4. We will compute a P-value for this test. α = 0.01.

5. Since nπ = 250(0.5) ≥ 10, and n(1 − π) = 250(0.5) ≥ 10, the large sample z test may be used.

6. n = 250, p = 140/250 = 0.56

$$z = \frac{0.56 - .5}{\sqrt{\dfrac{0.5(0.5)}{250}}} = \frac{.06}{.0316} = 1.90$$

7. P-value = 2(area under the z curve to the right of 1.90) ≈ 2(0.0287) = 0.0574

8. Since the P-value is greater than α of 0.01, H₀ is not rejected.

9. There is not enough evidence to suggest that the proportion of the time that the Belgium Euro coin would land with its head side up is not 0.5. With a significance level of 0.05, the same conclusion would be reached, since the p-value would still be greater than α.

10.33   a   1.   Let π represent the proportion of 18-19 year olds who have been asked to buy cigarettes for an underage smoker.

2. H₀: π = 0.50

3. H_a: $\pi < 0.50$

4. We will compute a P-value for this test. Even though the problem doesn't state what value is to be used for α, for illustrative purposes we will use $\alpha = 0.05$.

5. Since $n\pi = 149(0.5) \geq 10$, and $n(1 - \pi) = 149(0.5) \geq 10$, the large sample z test may be used.

6. $n = 149$, $p = 0.436$

$$z = \frac{0.436 - 0.5}{\sqrt{\dfrac{0.5(0.5)}{149}}} = \frac{-0.064}{0.04096} = -1.5624$$

7. P-value = area under the z curve to the left of $-1.5624 \approx 0.0591$

8. Since the P-value is greater than α, H_o cannot be rejected. However, if $\alpha = 0.10$ is used, then H_o will be rejected.

b   **1.** Let π represent the proportion of nonsmoking 18-19 year olds who have been asked to buy cigarettes for an underage smoker.

2. H_o: $\pi = 0.50$

3. H_a: $\pi < 0.50$

4. We will compute a P-value for this test. Even though the problem doesn't state what value is to be used for α, for illustrative purposes we will use $\alpha = 0.05$.

5. Since $n\pi = 110(0.5) \geq 10$, and $n(1 - \pi) = 110(0.5) \geq 10$, the large sample z test may be used.

6. $n = 110$, $p = 0.382$

$$z = \frac{0.382 - 0.5}{\sqrt{\dfrac{0.5(0.5)}{110}}} = \frac{-0.118}{0.04767} = -2.475$$

7. P-value = area under the z curve to the left of −2.475 ≈ 0.0067

8. Since the P-value is less than α, H_o can be rejected.

9. There is sufficient evidence (even with $\alpha = 0.01$) to support the claim that less than half of nonsmoking 18-19 year olds have been approached to buy cigarettes.

10.35 1. Let π represent the proportion of U.S. adults who see lottery or sweepstakes win as their best chance at accumulating $500,000.

2. H_o: $\pi = 0.25$

3. H_a: $\pi > 0.25$

4. $\alpha = 0.01$

5. Since $n\pi = 1010(0.25) \geq 10$, and $n(1 - \pi) = 1010(0.25) \geq 10$, the large sample z test may be used.

6. $n = 1010$, $p = 0.28$

$$z = \frac{0.28 - 0.25}{\sqrt{\dfrac{0.25(0.75)}{1010}}} = \frac{0.03}{0.01363} = 2.2018$$

7. P-value = area under the z curve to the right of $2.2018 = 0.0138$.

8. Since the P-value is greater than α, H_o cannot be rejected.

9. There is insufficient evidence (at $\alpha = 0.01$) to conclude that more than 25% of the U.S. adults see a lottery or sweepstakes win as their best chance at accumulating $500,000.

10.37 **a** 1. Let π represent the proportion of consumers who prefer fat fried chicken nuggets to baked chicken nuggets

2. H_o: $\pi = 0.6667$

3. H_a: $\pi > 0.6667$

4.	Even though the problem doesn't state what value is to be used for α, for illustrative purposes we will use $\alpha = 0.05$.

5.	Since $n\pi = 37(0.6667) \geq 10$, and $n(1 - \pi) = 37(0.3333) \geq 10$, the large sample z test may be used.

6.	$n = 37$, $p = 25/37 = 0.6757$

$$z = \frac{0.6757 - 0.6667}{\sqrt{\dfrac{0.6667(0.3333)}{37}}} = \frac{0.0090}{0.0775} = 0.1161$$

7.	P-value = area under the z curve to the right of $0.12 = 0.4522$.

8.	Since the P-value is greater than α, H_o cannot be rejected.

9.	There is insufficient evidence to conclude that more than 2/3 of consumers prefer fat fried chicken nuggets to baked chicken nuggets

b	A type I error is concluding that more than two-thirds of consumers prefer fat fried chicken nuggets to baked chicken nuggets, when in truth, less than two thirds prefer fat fried chicken nuggets.

c.	In part a, the conclusion was that H_o cannot be rejected. Therefore, if this was an error, it would be a Type II error. A type I error would be if the H_o was rejected in error. As the H_o wasn't rejected, a Type I error could not have been made.

d	1.	Let π represent the proportion of consumers who prefer beef steak fingers to hot-air fried beef steak fingers?

2.	H_o: $\pi = 0.5$

3.	H_a: $\pi > 0.5$

4.	Even though the problem doesn't state what value is to be used for α, for illustrative purposes we will use $\alpha = 0.05$.

5.	Since $n\pi = 37(0.5) \geq 10$, and $n(1 - \pi) = 37(0.5) \geq 10$, the large sample z test may be used.

6.	$n = 37$, $p = 23/37 = 0.6216$

$$z = \frac{0.6216 - 0.5}{\sqrt{\dfrac{0.5(0.5)}{37}}} = \frac{0.1216}{0.0822} = 1.48$$

7.   P-value = area under the z curve to the right of 1.48 = 0.0694.

8.   Since the P-value is greater than α, H_o cannot be rejected. There is insufficient evidence to conclude that consumers prefer beef steak fingers to hot-air fried beef steak fingers.

10.39   1.   Let π denote the proportion of all students in the population sampled that engaged in this form of cheating.

2. H_o: $\pi = 0.20$

3. H_a: $\pi > 0.20$

4. $\alpha = 0.05$ (A value for α was not specified in the problem, so this value was chosen.)

5.   Since $n\pi = 480(0.2) \geq 10$, and $n(1 - \pi) = 480(0.8) \geq 10$, the large sample z test may be used.
$$z = \frac{p - 0.20}{\sqrt{\dfrac{0.20(0.80)}{n}}}$$

6. $n = 480$, $x = 124$, $p = \dfrac{124}{480} = 0.258333$

$$z = \frac{0.258333 - 0.20}{\sqrt{\dfrac{0.20(0.80)}{480}}} = \frac{0.058333}{0.018257} = 3.195 \approx 3.20$$

7.   P-value = area under the z curve to the right of 3.20 = 1 − 0.9993 = 0.0007.

8.   Since the P-value is less than α, H_o is rejected.

9.   There is sufficient evidence in this sample to support the conclusion that the true proportion of students that engage in this form of cheating exceeds 0.20.

Exercises 10.41 – 10.58

10.41   Since this is a two-tailed test, the P-value is equal to twice the area captured in the tail in which z falls. Using Appendix Table II, the P-values are:

      a $2(0.0179) = 0.0358$

      b $2(0.0401) = 0.0802$

      c $2(0.2810) = 0.5620$

      d $2(0.0749) = 0.1498$

      e $2(0) = 0$

10.43   a   P-value = area under the 8 d.f. t curve to the right of $2.0 = 0.040$

      b   P-value = area under the 13 d.f. t curve to the right of $3.2 = 0.003$

      c   P-value = area under the 10 d.f. t curve to the left of -2.4
                        = area under the 10 d.f. t curve to the right of $2.4 = 0.019$

      d   P-value = area under the 21 d.f. t curve to the left of -4.2
                        = area under the 21 d.f. t curve to the right of $4.2 = 0.0002$

      e   P-value = 2(area under the 15 d.f. t curve to the right of 1.6) = 2 $(0.065) = 0.13$

      f   P-value = 2(area under the 15 d.f. t curve to the right of 1.6) = 2 $(0.065) = 0.13$

      g   P-value = 2(area under the 15 d.f. t curve to the right of 6.3) = $2(0) = 0$

10.45   The P-value for this test is equal to the area under the 14 d.f. t curve to the right of $3.2 = 0.003$.

      a $\alpha = 0.05$, reject H_o

      b $\alpha = 0.01$, reject H_o

      c $\alpha = 0.001$, fail to reject H_o

10.47 **a** P-value = 2(area under the 12 d.f. t curve to the right of 1.6) = 2(0.068) = 0.136. Since P-value > α, H_o is not rejected.

b P-value = 2(area under the 12 d.f. t curve to the left of –1.6)
= 2(area under the 12 d.f. t curve to the right of 1.6) = 2(0.068) = 0.136.
Since P-value > α, H_o is not rejected.

c P-value = 2(area under the 24 d.f. t curve to the left of –2.6)
= 2(area under the 24 d.f. t curve to the right of 2.6) = 2(0.008) = 0.016.
Since P-value > α, H_o is not rejected.

d P-value = 2(area under the 24 d.f. t curve to the left of –3.6)
= 2(area under the 24 d.f. t curve to the right of 3.6) = 2(0.001) = 0.002.
H_o would be rejected for any α > 0.002.

10.49 **1.** Let μ denote the true average "speaking up" value for Asian men.

2. H_o: μ = 10

3. H_a: μ < 10

4. α = 0.05   (A value for α was not specified in the problem, so this value was chosen.)

5. Since the sample is reasonable large (≥ 30) independent and randomly selected, it is reasonable to use the t test.

6. Test statistic: $t = \dfrac{\bar{x} - 10}{\dfrac{s}{\sqrt{n}}}$

7. Computations: n = 64, $\bar{x}$ = 8.75, s = 2.57,

$$t = \frac{8.75 - 10}{\dfrac{2.57}{\sqrt{64}}} = -3.89$$

8. P-value = area under the 63 d.f. t curve to the left of -3.89 ≈ 0.

9. Conclusion: Since the P-value is less than α, H_o is rejected. The data in this sample does support the conclusion that the average "speaking up" score for Asian men is smaller than 10.0.

10.51 **1.** Population characteristic of interest: μ = population mean MWAL

2. H_o: $\mu = 25$

3. H_a: $\mu > 25$

4. $\alpha = 0.05$

5. Test statistic: $t = \dfrac{\bar{x} - 25}{\dfrac{s}{\sqrt{n}}}$

6. Assumptions: This test requires a random sample and either a large sample size (generally $n \geq 30$) or a normal population distribution. Since the sample size is only 5, and a box plot of the data does not show perfect symmetry, does this seem reasonable? Based on the their understanding of MWAL values, the authors of the article thought it was reasonable to assume that the population distribution was approximately normal and based on their expert judgment, we will proceed, with caution, with the t-test.

7. Computations: $n = 5$, $\bar{x} = 27.54$, $s = 5.47$

$$t = \dfrac{27.54 - 25}{\dfrac{5.47}{\sqrt{5}}} = 1.0$$

8. P-value = area under the 4 d.f. t curve to the right of 1 = 0.187

9. Conclusion: Since the P-value is greater than α, the null hypothesis is not rejected at the 0.01 level. There is not enough evidence to suggest that the mean MWAL exceeds 25.

10.53 **1.** Let μ denote the mean rating given by all nonfundamentalists to Christian fundamentalists.

2. H_o: $\mu = 57$

3. H_a: $\mu < 57$

4. $\alpha = 0.01$ (A value for α was not specified in the problem. The value 0.01 was chosen for illustration.)

5. Since the sample is reasonable large (≥ 30) independent and randomly selected, it is reasonable to use the t test.

6. Test statistic: $t = \dfrac{\bar{x} - 57}{\dfrac{s}{\sqrt{n}}}$ with d.f. $= 960 - 1 = 959$

7. Computations: $n = 960$, $\bar{x} = 47$, $s = 21$

$$t = \frac{47 - 57}{\dfrac{21}{\sqrt{960}}} = \frac{-10}{0.6778} = -14.7542$$

8. P-value = area under the 959 d.f. t curve to the left of $-14.7542 \approx 0.0$

9. Since the P-value is smaller than α, H_0 is rejected. The sample data do provide convincing evidence that the average rating given by all nonfundamentalists to Christian fundamentalists is below 57.

10.55　**1.** Let μ denote the true average attention span (in minutes) of teenage Australian boys.

2. H_0: $\mu = 5$

3. H_a: $\mu < 5$

4. $\alpha = 0.01$

5. Since the sample is reasonable large (≥ 30) independent and randomly selected, it is reasonable to use the t test.

6. Test statistic: $t = \dfrac{\bar{x} - 5}{\dfrac{s}{\sqrt{n}}}$ with d.f. $= 50 - 1 = 49$

7. Computations: $n = 50$, $\bar{x} = 4$, $s = 1.4$

$$t = \frac{4 - 5}{\dfrac{1.4}{\sqrt{50}}} = \frac{-1}{0.198} = -5.0508$$

8. P-value = area under the 49 d.f. t curve to the left of $-5.0508 \approx 0.0$

9. Since the P-value is smaller than α, H$_o$ is rejected. The sample does provide convincing evidence that the average attention span of Australian teenagers is less than 5 minutes.

10.57 **a** Since the boxplot is nearly symmetric and the normal probability plot is very much like a straight line, it is reasonable to use a t-test to carry out the hypothesis test on μ.

b The median is slightly less than 245 and because of the near symmetry, the mean should be close to 245. Also, because of the large amount of variability in the data, it is quite conceivable that the average calorie content is 240.

c **1.** Let μ denote the true average calorie content of this type of frozen dinner.

2. H$_o$: $\mu = 240$

3. H$_a$: $\mu \neq 240$

4. $\alpha = 0.05$ (A value for α was not specified in the problem, so this value was chosen.)

5. Since the sample is reasonable large (≥ 30) independent and randomly selected, it is reasonable to use the t test.

6. Test statistic: $t = \dfrac{\overline{x} - 240}{\dfrac{s}{\sqrt{n}}}$ with d.f. = 12 – 1 = 11

7. Computations: n = 12, $\overline{x} = 244.333$, s = 12.383

$$t = \frac{244.333 - 240}{\dfrac{12.383}{\sqrt{12}}} = \frac{4.333}{3.575} = 1.21$$

8. P-value = 2(area under the 11 d.f. t curve to the right of 1.21) $\approx$ 2 (0.128) = 0.256

9. Since the P-value exceeds α, H_o is not rejected. The sample evidence does not support the conclusion that the average calorie content differs from 240.

Exercises 10.58 – 10.65

10.59 a When the significance level is held fixed, increasing the sample size will increase the power of a test.

b When the sample size is held fixed, increasing the significance level will increase the power of a test.

10.61 a α = area under the z curve to the left of $-1.28 = 0.1003$.

b The decision rule is reject H_o if $\bar{x} < 10 - 1.28(0.1) \Rightarrow \bar{x} < 9.872$.

$$z = \frac{9.872 - 9.8}{0.1} = 0.72$$

(β when $\mu = 9.8$) = area under the z curve to the right of 0.72
$\phantom{(\beta \text{ when } \mu = 9.8)}$ = 1 – area under the z curve to the left of 0.72
$\phantom{(\beta \text{ when } \mu = 9.8)}$ = 1 – 0.7642 = 0.2358.

This means that if $\mu = 9.8$, about 24% of all samples would result in $\bar{x}$ values greater than 9.872 and the nonrejection of $H_o : \mu = 10$.

c β when $\mu = 9.5$ would be smaller than β when $\mu = 9.8$.

$$z = \frac{9.872 - 9.5}{0.1} = 3.72$$

(β when $\mu = 9.5$) = 1– area under the z curve to the left of 3.72 $\approx 1 - 1 = 0$

d When $\mu = 9.8$, the value of β is 0.2358 from part **b**. So, power of the test = 1-β = 1-0.2358 = 0.7642. When $\mu = 9.5$, the value of β is (approximately) 0 from part **c**. So, power of the test = 1-β = 1.

10.63 a **1.** Let μ denote the mean amount of shaft wear after a fixed mileage.

2. $H_o: \mu = 0.035$

3. H_a: $\mu > 0.035$

4. $\alpha = 0.05$

5. $t = \dfrac{\bar{x} - 0.035}{\dfrac{s}{\sqrt{n}}}$ with d.f. = 6

6. $t = \dfrac{0.0372 - 0.035}{\dfrac{0.0125}{\sqrt{7}}} = 0.466$

7. P-value = area under the 6 d.f. t curve to the right of 0.466 ≈ 0.329.

8. Since the P-value exceeds α, H_o is not rejected. It cannot be concluded that the mean amount of shaft wear exceeds 0.035 inches.

b $d = \dfrac{|\text{alternative value - hypotheszed value}|}{\sigma} = \dfrac{|0.04 - 0.035|}{0.0125} = 0.4$

From Appendix Table V, use the set of curves for $\alpha = 0.05$, one-tailed test. Enter the table using d = 0.4, and go up to where the d = 0.4 line intersects with the curve for d.f. = 6. Then read the β value of the vertical axis. This leads to a β of about 0.75.

c   From **b**, value of β is about 0.75. So, power of the test = 1-β $\approx$ 1-0.75 = 0.25.

10.65   **a** $d = \dfrac{|0.52 - 0.5|}{0.02} = 1$ From Appendix Table V, β $\approx$0.06.

b $d = \dfrac{|0.48 - 0.5|}{0.02} = 1$ From Appendix Table V, β $\approx$0.06.

c $d = \dfrac{|0.52 - 0.5|}{0.02} = 1$ From Appendix Table V, β $\approx$0.21

d $d = \dfrac{|0.54 - 0.5|}{0.02} = 2$ From Appendix Table V, β $\approx$ 0.

e $d = \dfrac{|0.54 - 0.5|}{0.04} = 1$ From Appendix Table V, $\beta \approx 0.06$.

f $d = \dfrac{|0.54 - 0.5|}{0.04} = 1$ From Appendix Table V, $\beta \approx 0.01$.

g   Answers will vary from student to student.

Exercises 10.66 – 10.79

10.67   **1.**   Let π represent the proportion of job applicants in California that test positive for drug use.

2. H_o: $\pi = 0.10$

3. H_a: $\pi > 0.10$

4. $\alpha = 0.01$ (A value for α was not specified in the problem, so this value was chosen.)

5.   Since $n\pi = 600(0.10) = 60 \geq 10$, and $n(1 - \pi) = 600(0.90) = 540 \geq 10$, the large sample z test may be used.

6. $n = 600$, $x = 73$, $p = \dfrac{73}{600} = 0.1217$

$$z = \frac{0.1217 - 0.1}{\sqrt{\dfrac{0.1(0.9)}{600}}} = \frac{0.0217}{0.0122} = 1.78$$

7.   P-value = area under the z curve to the right of $1.78 = 1 - 0.9625 = 0.0375$.

8.   Since the P-value exceeds α, H_o is not rejected.

9.   The data does not support the conclusion that the proportion of job applicants in California that test positive for drug use exceeds 0.10.

10.69   **1.**   Let π denote the true proportion of all cars purchased in this area that were white.

2. H_o: $\pi = 0.20$

3. $H_a: \pi \neq 0.20$

4. See # 8 below

5. Since $n\pi = 400(0.2) \geq 10$, and $n(1 - \pi) = 400(0.8) \geq 10$, the large sample z test may be used.

6. $n = 400$, $x = 100$, $p = 0.25$

$$z = \frac{0.25 - 0.20}{\sqrt{\dfrac{0.2(0.8)}{400}}} = 2.50$$

7. P-value = 2(area under the z curve to the right of 2.50) = $2(1 - 0.9938) = 0.0124$.

8. With $\alpha = 0.05$, the P-value is less than α and the null hypothesis would be rejected. The conclusion would be that the true proportion of cars sold that are white differs from the national rate of 0.20.

With $\alpha = 0.01$, the P-value is greater than α and the null hypothesis would not be rejected. Then the sample does not support the conclusion that the true proportion of cars sold that are white differs from 0.20.

10.71 Let μ denote the true average time to change (in months).

$H_0: \mu = 24$
$H_a: \mu > 24$
$\alpha = 0.01$

Test statistic: $t = \dfrac{\bar{x} - 24}{\dfrac{s}{\sqrt{n}}}$

Computations: $n = 44$, $\bar{x} = 35.02$, $s = 18.94$,

$$t = \frac{35.02 - 24}{\dfrac{18.94}{\sqrt{44}}} = 3.86$$

P-value = area under the 43 d.f. t curve to the right of $3.86 \approx 1 - 1 = 0$.

Since the P-value is less than α, H_o is rejected. There is sufficient information in this sample to support the conclusion that the true average time to change exceeds two years.

10.73 **a** Daily caffeine consumption cannot be a negative value. Since the standard deviation is larger than the mean, this would imply that a sizable portion of a normal curve with this mean and this standard deviation would extend into the negative values on the number line. Therefore, it is not plausible that the population distribution of daily caffeine consumption is normal.

Since the sample size is large (greater than 30) the Central Limit Theorem allows for the conclusion that the distribution of $\bar{x}$ is approximately normal even though the population distribution is not normal. So it is not necessary to assume that the population distribution of daily caffeine consumption is normal to test hypotheses about the value of population mean consumption.

b Let μ denote the population mean daily consumption of caffeine.

$H_o : \mu = 200$
$H_a : \mu > 200$
$\alpha = 0.10$

$$t = \frac{\bar{x} - 200}{\dfrac{s}{\sqrt{n}}}$$

$n = 47$, $\bar{x} = 215$, $s = 235$

$$t = \frac{\bar{x} - 200}{\dfrac{s}{\sqrt{n}}} = \frac{215 - 200}{\dfrac{235}{\sqrt{47}}} = 0.44$$

P-value = area under the 46 d.f. t curve to the right of 0.44 = 1 − 0.6700 = 0.33

Since the P-value exceeds the level of significance of 0.10, H_o is not rejected. The data does not support the conclusion that the population mean daily caffeine consumption exceeds 200 mg.

10.75 **1.** Population characteristic of interest: μ = true average fuel efficiency

2. $H_o: \mu = 30$

187

3. $H_a: \mu < 30$

4. $\alpha = 0.05$ (for demonstration purposes)

5. Test statistic: $t = \dfrac{\bar{x} - 30}{\dfrac{s}{\sqrt{n}}}$

6. Assumptions: This test requires a random sample and either a large sample size (generally $n \geq 30$) or a normal population distribution. Since the sample size is only 6, we can look at a box plot of the data. It shows symmetry, indicating that it would not be unreasonable to assume that the population would be approximately normal. Hence we can proceed with a t-test.

7. Computations: $n = 6$, $\bar{x} = 29.33$, $s = 1.41$

$$t = \frac{29.33 - 30}{\dfrac{1.41}{\sqrt{6}}} = -1.164$$

8. P-value = area under the 5 d.f. t curve to the left of -1.164 = 0.15

9. Conclusion: Since the P-value is greater than α, the null hypothesis is not rejected at the 0.01 level. The data does not contradict the prior belief that the true average fuel efficiency is at least 30.

10.77 Let μ denote the true mean time required to achieve 100° F with the heating equipment of this manufacturer.

$H_o: \mu = 15$
$H_a: \mu > 15$
$\alpha = 0.05$

The test statistic is: $t = \dfrac{\bar{x} - 15}{\dfrac{s}{\sqrt{n}}}$.

From the sample: $n = 25$, $\bar{x} = 17.5$, $s = 2.2$,

$$t = \frac{17.5 - 15}{\frac{2.2}{\sqrt{25}}} = 5.68.$$

P-value =area under the 24 d.f. t curve to the right of $5.68 \approx 1 - 1 = 0$. Because the P-value is smaller than α, H_o is rejected. The data does cast doubt on the company's claim that it requires at most 15 minutes to achieve 100° F.

10.79   P(Type I error) = $P(\alpha) = 0$
P(Type II error) = $P(\beta) = 0.1 \le P(\beta) \le 0.3$

Chapter 11

Exercises 11.1 – 11.28

11.1 $\mu_{\bar{x}_1 - \bar{x}_2} = \mu_1 - \mu_2 = 30 - 25 = 5$

$$\sigma_{\bar{x}_1 - \bar{x}_2} = \sqrt{\frac{\sigma_1^2}{n_1} + \frac{\sigma_2^2}{n_2}} = \sqrt{\frac{(2)^2}{40} + \frac{(3)^2}{50}} = \sqrt{\frac{4}{40} + \frac{9}{50}} = \sqrt{0.28} = 0.529$$

Since both n_1 and n_2 are large, the sampling distribution of $\bar{x}_1 - \bar{x}_2$ is approximately normal. It is centered at 5 and the standard deviation is 0.529.

11.3 μ_1 be the mean math anxiety for male at-risk students
. μ_2 be the mean math anxiety for female at-risk students

$H_0: \mu_1 - \mu_2 = 0$ $H_a: \mu_1 - \mu_2 \neq 0$

$\alpha = 0.05$

Test statistic: $t = \dfrac{(\bar{x}_1 - \bar{x}_2) - 0}{\sqrt{\dfrac{s_1^2}{n_1} + \dfrac{s_2^2}{n_2}}}$

Assumptions: The population distributions are approximately normal or the sample size is large (generally ≥ 30) and the two samples are independently selected random samples. Since one of the samples is less than 30 and we don't have access to the raw sample data, we must assume that the population distributions are approximately normal.

$n_1 = 20$, $\bar{x}_1 = 35.9$, $s_1 = 11.9$, $n_2 = 38$, $\bar{x}_2 = 36.6$, $s_2 = 12.3$

$$t = \frac{(35.9 - 36.6) - 0}{\sqrt{\dfrac{(11.9)^2}{20} + \dfrac{(12.3)^2}{38}}} = \frac{-0.7}{3.326} = -0.21$$

$$df = \frac{\left(\dfrac{s_1^2}{n_1} + \dfrac{s_2^2}{n_2}\right)^2}{\dfrac{1}{n_1 - 1}\left(\dfrac{s_1^2}{n_1}\right)^2 + \dfrac{1}{n_2 - 1}\left(\dfrac{s_2^2}{n_2}\right)^2} = \frac{(7.081 + 3.981)^2}{\dfrac{(7.081)^2}{19} + \dfrac{(3.981)^2}{37}} = 39.89$$

So $df = 39$ (rounded down to an integer)

P-value = 2(the area under the 39 df t curve to the left of –0.21 $\approx$ 2(0.421) = 0.842.

Since the P-value is greater than α, the null hypothesis is not rejected at the 0.05 level of significance. There is not enough evidence to show that mean math anxiety score of at risk males differs from that of at-risk females.

11.5　　a　　To use a 2 sample t- test, one of the requirements is the samples are either large (generally $\geq$ 30) or that the population distributions are approximately normally distributed. In this case both sample sizes are small (10) and we do not any information about the population distributions.

　　　　b　　No the sample sizes are large enough that we can, with the use of the Central Limit Theorem, assume that the sampling distribution of $\bar{x}_1 - \bar{x}_2$ is approximately normal.

　　　　c　　μ_1 be the mean fumonisin level for partially degermed corn meal
　　　　　　μ_2 be the mean fumonisin level for partially degermed corn meal
　　　　　　H_0: $\mu_1 - \mu_2 = 0$ 　　H_a: $\mu_1 - \mu_2 \neq 0$

　　　　　　$\alpha = 0.01$

　　　　　　Test statistic:　　$t = \dfrac{(\bar{x}_1 - \bar{x}_2) - 0}{\sqrt{\dfrac{s_1^2}{n_1} + \dfrac{s_2^2}{n_2}}}$

　　　　　　Assumptions: The population distributions are approximately normal or the sample size is large (generally $\geq$ 30) and the two samples are independently selected random samples.

　　　　　　$n_1 = 50$, $\bar{x}_1 = 0.59$, $s_1 = 1.01$, $n_2 = 50$, $\bar{x}_2 = 1.21$, $s_2 = 1.71$

　　　　　　$t = \dfrac{(0.59 - 1.21) - 0}{\sqrt{\dfrac{(1.01)^2}{50} + \dfrac{(1.71)^2}{50}}} = \dfrac{-0.62}{0.2809} = -2.207$

$$df = \dfrac{\left(\dfrac{s_1^2}{n_1} + \dfrac{s_2^2}{n_2}\right)^2}{\dfrac{1}{n_1-1}\left(\dfrac{s_1^2}{n_1}\right)^2 + \dfrac{1}{n_2-1}\left(\dfrac{s_2^2}{n_2}\right)^2} = \dfrac{(0.0204 + 0.0585)^2}{\dfrac{(0.0204)^2}{49} + \dfrac{(0.0585)^2}{49}} = 79.47$$

So $df = 79$ (rounded down to an integer)

P-value = 2(the area under the 79 df t curve to the left of -2.207) ≈ 0.030

Since the P-value is greater than α, the null hypothesis is not rejected at the 0.01 level of significance. There is not enough evidence to show that there is a difference in mean fumonisin level for the two types of corn meal.

11.7 a Small prey:

Let μ_1 be the mean amount (mg) of venom injected by the inexperienced snakes and μ_2 the mean amount of venom injected by the experienced snakes when the prey is a small prey.

H_0: $\mu_1 - \mu_2 = 0$ H_a: $\mu_1 - \mu_2 \neq 0$

$\alpha = 0.05$ (A value for α is not specified in the problem. We use $\alpha=0.05$ for illustration.)

Test statistic: $t = \dfrac{(\bar{x}_1 - \bar{x}_2) - 0}{\sqrt{\dfrac{s_1^2}{n_1} + \dfrac{s_2^2}{n_2}}}$

Assumptions: The population distributions are (at least approximately) normal and the two samples are independently selected random samples.

$n_1 = 7$, $\bar{x}_1 = 3.1$, $s_1 = 1.0$, $n_2 = 7$, $\bar{x}_2 = 2.6$, $s_2 = 0.3$

$$t = \dfrac{(3.1 - 2.6) - 0}{\sqrt{\dfrac{(1.0)^2}{7} + \dfrac{(0.3)^2}{7}}} = \dfrac{0.5}{0.3946} = 1.2670$$

$$df = \frac{\left(\dfrac{s_1^2}{n_1} + \dfrac{s_2^2}{n_2}\right)^2}{\dfrac{1}{n_1-1}\left(\dfrac{s_1^2}{n_1}\right)^2 + \dfrac{1}{n_2-1}\left(\dfrac{s_2^2}{n_2}\right)^2} = \frac{(0.1428 + 0.0128)^2}{\dfrac{(0.1428)^2}{6} + \dfrac{(0.0128)^2}{6}} = 7.0713$$

So $df = 7$ (rounded down to an integer)

P-value = 2 times the area under the 7 df t curve to the right of $1.2670 \approx 0.2456$.

Since the P-value is greater than α, the null hypothesis cannot be rejected. At level of significance of 0.05 (or even 0.10), the data do not indicate that there is a difference in the amount of venom injected between inexperienced snakes and experienced snakes when the prey is a small prey.

b   Medium prey:

Let μ_1 be the mean amount (mg) of venom injected by the inexperienced snakes and μ_2 the mean amount of venom injected by the experienced snakes for medium prey.

H_0: $\mu_1 - \mu_2 = 0$ H_a: $\mu_1 - \mu_2 \neq 0$

$\alpha = 0.05$ (A value for α is not specified in the problem. We use α=0.05 for illustration.)

Test statistic: $t = \dfrac{(\bar{x}_1 - \bar{x}_2) - 0}{\sqrt{\dfrac{s_1^2}{n_1} + \dfrac{s_2^2}{n_2}}}$

Assumptions: The population distributions are (at least approximately) normal and the two samples are independently selected random samples.

$n_1 = 7$, $\bar{x}_1 = 3.4$, $s_1 = 0.4$, $n_2 = 7$, $\bar{x}_2 = 2.9$, $s_2 = 0.6$

$t = \dfrac{(3.4 - 2.9) - 0}{\sqrt{\dfrac{(0.4)^2}{7} + \dfrac{(0.6)^2}{7}}} = \dfrac{0.5}{0.2725} = 1.8344$

$$df = \frac{\left(\dfrac{s_1^2}{n_1} + \dfrac{s_2^2}{n_2}\right)^2}{\dfrac{1}{n_1-1}\left(\dfrac{s_1^2}{n_1}\right)^2 + \dfrac{1}{n_2-1}\left(\dfrac{s_2^2}{n_2}\right)^2} = \frac{(0.0228 + 0.0514)^2}{\dfrac{(0.0228)^2}{6} + \dfrac{(0.0514)^2}{6}} = 10.45$$

So $df = 10$ (rounded down to an integer)

P-value = 2 times the area under the 10 df t curve to the right of $1.8344 \approx 0.0964$.

Since the P-value is greater than α, the null hypothesis cannot be rejected. At level of significance 0.05, the data do not indicate that there is a difference in the amount of venom injected between inexperienced snakes and experienced snakes for medium prey.

c   Large prey:

Let μ_1 be the mean amount (mg) of venom injected by the inexperienced snakes and μ_2 the mean amount of venom injected by the experienced snakes for large prey.

H$_0$: $\mu_1 - \mu_2 = 0$ H$_a$: $\mu_1 - \mu_2 \neq 0$

$\alpha = 0.05$ (A value for α is not specified in the problem. We use $\alpha=0.05$ for illustration.)

Test statistic: $t = \dfrac{(\bar{x}_1 - \bar{x}_2) - 0}{\sqrt{\dfrac{s_1^2}{n_1} + \dfrac{s_2^2}{n_2}}}$

Assumptions: The population distributions are (at least approximately) normal and the two samples are independently selected random samples.

$n_1 = 7$, $\bar{x}_1 = 1.8$, $s_1 = 0.3$, $n_2 = 7$, $\bar{x}_2 = 4.7$, $s_2 = 0.3$

$$t = \frac{(1.8 - 4.7) - 0}{\sqrt{\dfrac{(0.3)^2}{7} + \dfrac{(0.3)^2}{7}}} = \frac{-2.9}{0.1603} = -18.0846$$

$$df = \dfrac{\left(\dfrac{s_1^2}{n_1} + \dfrac{s_2^2}{n_2}\right)^2}{\dfrac{1}{n_1-1}\left(\dfrac{s_1^2}{n_1}\right)^2 + \dfrac{1}{n_2-1}\left(\dfrac{s_2^2}{n_2}\right)^2} = \dfrac{(0.0128 + 0.0128)^2}{\dfrac{(0.0128)^2}{6} + \dfrac{(0.0128)^2}{6}} = 12.0. \qquad \text{So } df = 12$$

P-value = 2 times the area under the 12 df t curve to the left of -18.0846 ≈ 0.0000.

Since the P-value is smaller than α, the null hypothesis is rejected. The data provide strong evidence that there is a difference in the amount of venom injected between inexperienced snakes and experienced snakes for large prey.

11.9 **a** Let μ_1 denote the true mean level of testosterone for male trial lawyers and μ_2 the true mean level of testosterone for male nontrial lawyers. We wish to test the null hypothesis H₀: $\mu_1 - \mu_2 = 0$ against the alternative hypothesis Hₐ: $\mu_1 - \mu_2 \neq 0$. The t statistic for this test is reported to be 3.75 and the degrees of freedom are 64. The P-value is twice the area under the 64 df t curve to the right of 3.75. This is equal to 0.0004 (the report states that the P-value is < 0.001 but doesn't report the actual P-value). Hence the data do provide strong evidence to conclude that the mean testosterone levels for male trial lawyers and nontrial lawyers are different.

b Let μ_1 denote the true mean level of testosterone for female trial lawyers and μ_2 the true mean level of testosterone for female nontrial lawyers. We wish to test the null hypothesis H₀: $\mu_1 - \mu_2 = 0$ against the alternative hypothesis Hₐ: $\mu_1 - \mu_2 \neq 0$. The t-statistic for this test is reported to be 2.26 and the degrees of freedom are 29. The P-value is twice the area under the 29 df t curve to the right of 2.26. This is equal to 0.0316 (the report states that the P-value is < 0.05 but doesn't report the actual P-value). Hence the data do provide sufficient evidence to conclude that the mean testosterone levels for female trial lawyers and female nontrial lawyers are different.

c There is not enough information to carry out a test to determine whether there is a significant difference in the mean testosterone levels of male and female trial lawyers. To carry out such a test we need the sample means and sample standard deviations for the 35 male trial lawyers and the 13 female trial lawyers.

11.11 **a** Let μ_1 be the true mean "appropriateness" score assigned to wearing a hat in a class by the population of students and μ_2 be the corresponding score for faculty.

H₀: $\mu_1 - \mu_2 = 0$ Hₐ: $\mu_1 - \mu_2 \neq 0$

$\alpha = 0.05$ (A value for α is not specified in the problem. We use $\alpha=0.05$ for illustration.)

Test statistic: $\quad t = \dfrac{(\bar{x}_1 - \bar{x}_2) - 0}{\sqrt{\dfrac{s_1^2}{n_1} + \dfrac{s_2^2}{n_2}}}$

Assumptions: The sample sizes for the two groups are large (say, greater than 30 for each) and the two samples are independently selected random samples.

$n_1 = 173,\ \bar{x}_1 = 2.80,\ s_1 = 1.0,\ n_2 = 98,\ \bar{x}_2 = 3.63,\ s_2 = 1.0$

$$t = \frac{(2.80 - 3.63) - 0}{\sqrt{\dfrac{(1.0)^2}{173} + \dfrac{(1.0)^2}{98}}} = \frac{-0.83}{0.1264} = -6.5649$$

$$df = \frac{\left(\dfrac{s_1^2}{n_1} + \dfrac{s_2^2}{n_2}\right)^2}{\dfrac{1}{n_1 - 1}\left(\dfrac{s_1^2}{n_1}\right)^2 + \dfrac{1}{n_2 - 1}\left(\dfrac{s_2^2}{n_2}\right)^2} = \frac{(0.00578 + 0.01020)^2}{\dfrac{(0.00578)^2}{172} + \dfrac{(0.01020)^2}{97}} = 201.5$$

So $df = 201$ (rounded down to an integer)

P-value = 2 times the area under the 201 df t curve to the left of $-6.5649 \approx 0.0000$.

Since the P-value is much less than α, the null hypothesis of no difference is rejected. The data do provide very strong evidence to indicate that there is a difference in the mean appropriateness scores between students and faculty for wearing hats in the class room. The mean appropriateness score for students is significantly smaller than that for faculty.

b   Let μ_1 be the true mean "appropriateness" score assigned to addressing an instructor by his or her first name by the population of students and μ_2 be the corresponding score for faculty.

H$_0$: $\mu_1 - \mu_2 {}_2 = 0$ H$_a$: $\mu_1 - \mu_2 > 0$

$\alpha = 0.05$ (A value for α is not specified in the problem. We use $\alpha=0.05$ for illustration.)

Test statistic: $t = \dfrac{(\bar{x}_1 - \bar{x}_2) - 0}{\sqrt{\dfrac{s_1^2}{n_1} + \dfrac{s_2^2}{n_2}}}$

Assumptions: The sample sizes for the two groups are large (say, greater than 30 for each) and the two samples are independently selected random samples.

$n_1 = 173$, $\bar{x}_1 = 2.90$, $s_1 = 1.0$, $n_2 = 98$, $\bar{x}_2 = 2.11$, $s_2 = 1.0$

$t = \dfrac{(2.90 - 2.11) - 0}{\sqrt{\dfrac{(1.0)^2}{173} + \dfrac{(1.0)^2}{98}}} = \dfrac{0.79}{0.1264} = 6.2485$

$df = \dfrac{\left(\dfrac{s_1^2}{n_1} + \dfrac{s_2^2}{n_2}\right)^2}{\dfrac{1}{n_1 - 1}\left(\dfrac{s_1^2}{n_1}\right)^2 + \dfrac{1}{n_2 - 1}\left(\dfrac{s_2^2}{n_2}\right)^2} = \dfrac{(0.00578 + 0.01020)^2}{\dfrac{(0.00578)^2}{172} + \dfrac{(0.01020)^2}{97}} = 201.5$

So $df = 201$ (rounded down to an integer)

P-value = the area under the 201 df t curve to the right of 6.2485 ≈ 0.0000.

Since the P-value is much less than α, the null hypothesis of no difference is rejected. The data do provide very strong evidence to indicate that the mean appropriateness score for addressing the instructor by his or her first name is higher for students than for faculty.

c   Let μ_1 be the true mean "appropriateness" score assigned to talking on a cell phone during class by the population of students and μ_2 be the corresponding score for faculty.

H_0: $\mu_1 - \mu_2 = 0$ H_a: $\mu_1 - \mu_2 \neq 0$

$\alpha = 0.05$ (A value for α is not specified in the problem. We use $\alpha = 0.05$ for illustration.)

Test statistic: $t = \dfrac{(\bar{x}_1 - \bar{x}_2) - 0}{\sqrt{\dfrac{s_1^2}{n_1} + \dfrac{s_2^2}{n_2}}}$

Assumptions: The sample sizes for the two groups are large (say, greater than 30 for each) and the two samples are independently selected random samples.

$$n_1 = 173, \ \bar{x}_1 = 1.11, \ s_1 = 1.0, \ n_2 = 98, \ \bar{x}_2 = 1.10, \ s_2 = 1.0$$

$$t = \frac{(1.11-1.10)-0}{\sqrt{\dfrac{(1.0)^2}{173}+\dfrac{(1.0)^2}{98}}} = \frac{-0.01}{0.1264} = 0.0791$$

$$df = \frac{\left(\dfrac{s_1^2}{n_1}+\dfrac{s_2^2}{n_2}\right)^2}{\dfrac{1}{n_1-1}\left(\dfrac{s_1^2}{n_1}\right)^2+\dfrac{1}{n_2-1}\left(\dfrac{s_2^2}{n_2}\right)^2} = \frac{(0.00578+0.01020)^2}{\dfrac{(0.00578)^2}{172}+\dfrac{(0.01020)^2}{97}} = 201.5$$

So $df = 201$ (rounded down to an integer)

P-value = 2 times the area under the 201 df t curve to the right of $0.0791 \approx 0.9370$.

Since the P-value is not less than α, the null hypothesis of no difference cannot be rejected. The data do not provide evidence to indicate that there is a difference in the mean appropriateness scores between students and faculty for talking on cell phones in class. The result does not imply that students and faculty consider it acceptable to talk on a cell phone during class. It simply says that data do not provide enough evidence to claim a difference exists.

11.13   **a**   Let μ_1 be the true mean stream gradient (%) for the population of sites with tailed frogs and μ_2 be the corresponding mean for sites without tailed frogs.

$H_0: \mu_1 - \mu_2 = 0$ $H_a: \mu_1 - \mu_2 \neq 0$

$\alpha = 0.01$

Test statistic: $t = \dfrac{(\bar{x}_1 - \bar{x}_2)-0}{\sqrt{\dfrac{s_1^2}{n_1}+\dfrac{s_2^2}{n_2}}}$

Assumptions: The distribution of stream gradients is approximately normal for both types of sites and the two samples are independently selected random samples.

$n_1 = 18$, $\bar{x}_1 = 9.1$, $s_1 = 6.0$, $n_2 = 31$, $\bar{x}_2 = 5.9$, $s_2 = 6.29$

$$t = \frac{(9.1 - 5.9) - 0}{\sqrt{\frac{(6.00)^2}{18} + \frac{(6.29)^2}{31}}} = \frac{3.2}{1.8100} = 1.7679$$

$$df = \frac{\left(\frac{s_1^2}{n_1} + \frac{s_2^2}{n_2}\right)^2}{\frac{1}{n_1 - 1}\left(\frac{s_1^2}{n_1}\right)^2 + \frac{1}{n_2 - 1}\left(\frac{s_2^2}{n_2}\right)^2} = \frac{(2.00 + 1.2763)^2}{\frac{(2.00)^2}{17} + \frac{(1.2763)^2}{30}} = 37.07$$

So $df = 37$ (rounded down to an integer)

P-value = 2 times the area under the 37 df t curve to the right of 1.7679 $\approx$ 0.0853.

Since the P-value is greater than α, the null hypothesis of no difference cannot be rejected. The data do not provide sufficient evidence to suggest a difference between the mean stream gradients for sites with tailed frogs and sites without tailed frogs.

b   Let μ_1 be the true mean water temperature for sites with tailed frogs and μ_2 be the corresponding mean for sites without tailed frogs.

H$_0$: $\mu_1 - \mu_2 = 0$ H$_a$: $\mu_1 - \mu_2 \neq 0$

$\alpha = 0.01$

Test statistic: $t = \dfrac{(\bar{x}_1 - \bar{x}_2) - 0}{\sqrt{\dfrac{s_1^2}{n_1} + \dfrac{s_2^2}{n_2}}}$

Assumptions: The distribution of stream gradients is approximately normal for both types of sites and the two samples are independently selected random samples.

$n_1 = 18$, $\bar{x}_1 = 12.2$, $s_1 = 1.71$, $n_2 = 31$, $\bar{x}_2 = 12.8$, $s_2 = 1.33$

$$t = \frac{(12.2 - 12.8) - 0}{\sqrt{\frac{(1.71)^2}{18} + \frac{(1.33)^2}{31}}} = \frac{-0.6}{0.4685} = -1.2806$$

200

$$df = \frac{\left(\dfrac{s_1^2}{n_1} + \dfrac{s_2^2}{n_2}\right)^2}{\dfrac{1}{n_1-1}\left(\dfrac{s_1^2}{n_1}\right)^2 + \dfrac{1}{n_2-1}\left(\dfrac{s_2^2}{n_2}\right)^2} = \frac{(0.1625+0.0571)^2}{\dfrac{(0.1625)^2}{17} + \dfrac{(0.0571)^2}{30}} = 29.01$$

So *df* = 29 (rounded down to an integer)

P-value = 2 times the area under the 29 df t curve to the left of -1.2806 ≈ 0.2105.

Since the P-value is greater than α, the null hypothesis of no difference cannot be rejected. The data do not provide sufficient evidence to conclude that the mean water temperatures for the two types of sites (with and without tailed frogs) are different.

c   Let μ_1 be the true mean stream depth for the population of sites with tailed frogs and μ_2 be the corresponding mean for sites without tailed frogs.

H$_0$: $\mu_1 - \mu_2 = 0$ H$_a$: $\mu_1 - \mu_2 \neq 0$

$\alpha = 0.01$

Test statistic: $t = \dfrac{(\bar{x}_1 - \bar{x}_2) - 0}{\sqrt{\dfrac{s_1^2}{n_1} + \dfrac{s_2^2}{n_2}}}$

Assumptions: The sample sizes for each group is large (greater than or equal to 30) and the two samples are independently selected random samples. Since the sample sizes for the two samples are 82 and 267 respectively, it is quite reasonable to use the independent samples t test for comparing the mean depths for the two types of sites.

$n_1 = 82$, $\bar{x}_1 = 5.32$, $s_1 = 2.27$, $n_2 = 267$, $\bar{x}_2 = 8.46$, $s_2 = 5.95$

$$t = \frac{(5.32 - 8.46) - 0}{\sqrt{\dfrac{(2.27)^2}{82} + \dfrac{(5.95)^2}{267}}} = \frac{-3.14}{0.4421} = -7.1028$$

$$df = \dfrac{\left(\dfrac{s_1^2}{n_1} + \dfrac{s_2^2}{n_2}\right)^2}{\dfrac{1}{n_1 - 1}\left(\dfrac{s_1^2}{n_1}\right)^2 + \dfrac{1}{n_2 - 1}\left(\dfrac{s_2^2}{n_2}\right)^2} = \dfrac{(0.06284 + 0.13259)^2}{\dfrac{(0.06284)^2}{81} + \dfrac{(0.13259)^2}{266}} = 332.6$$

So $df = 332$ (rounded down to an integer)

P-value = 2 times the area under the 332 df t curve to the left of -7.1028 ≈ 0.0000.

Since the P-value is much smaller than α, the null hypothesis of no difference is rejected. The data provide very strong evidence to conclude that there indeed is a difference between the mean stream depths of sites with tailed frogs and sites without tailed frogs.

11.15 **a** Let μ denote the mean salary (in Canadian dollars) for the population of female MBA graduates of this Canadian business school.

H_0: $\mu = 100,000$ H_a: $\mu > 100,000$

A value for α was not specified in the problem. We will compute the P-value.

Test statistic: $t = \dfrac{\bar{x} - 100,000}{\dfrac{s}{\sqrt{n}}}$ with d.f. = 233 – 1 = 232

Computations: n = 233, $\bar{x}$ = 105,156, s = 98,525

$$t = \dfrac{105,156 - 100,000}{\dfrac{98,525}{\sqrt{233}}} = \dfrac{5156.0}{6454.587} = 0.7988$$

P-value = area under the 232 d.f. t curve to the right of 0.7988 ≈ 0.2126

For significance levels greater than 0.2126 we can conclude that the mean salary of female MBA graduates from this business school is above 100,000 dollars.

b Let μ_1 be the true mean salary for female MBA graduates from this business school and μ_2 be the mean for male MBA graduates.

H_0: $\mu_1 - \mu_2 = 0$ H_a: $\mu_1 - \mu_2 < 0$

$\alpha = 0.01$ (a value for α is not specified in this problem. We will use $\alpha = 0.01$ for illustration.)

Test statistic: $t = \dfrac{(\bar{x}_1 - \bar{x}_2) - 0}{\sqrt{\dfrac{s_1^2}{n_1} + \dfrac{s_2^2}{n_2}}}$

Assumptions: The sample sizes for each group is large (greater than or equal to 30) and the two samples are independently selected random samples.

$n_1 = 233$, $\bar{x}_1 = 105{,}156$, $s_1 = 98{,}525$, $n_2 = 258$, $\bar{x}_2 = 133{,}442$, $s_2 = 131{,}090$

$t = \dfrac{(105{,}156 - 133{,}442) - 0}{\sqrt{\dfrac{(98{,}525)^2}{233} + \dfrac{(131{,}090)^2}{258}}} = \dfrac{-28{,}286}{10405.22} = -2.718$

$df = \dfrac{\left(\dfrac{s_1^2}{n_1} + \dfrac{s_2^2}{n_2}\right)^2}{\dfrac{1}{n_1 - 1}\left(\dfrac{s_1^2}{n_1}\right)^2 + \dfrac{1}{n_2 - 1}\left(\dfrac{s_2^2}{n_2}\right)^2} = \dfrac{(41{,}661{,}697.96 + 66{,}606{,}930.62)^2}{\dfrac{(41{,}661{,}697.96)^2}{232} + \dfrac{(66{,}606{,}930.62)^2}{257}} = 473.7$

So $df = 473$ (rounded down to an integer)

P-value = the area under the 473 df t curve to the left of $-2.718 \approx 0.0034$.

Since the P-value is much smaller than α, the null hypothesis of no difference is rejected. The data provide very strong evidence to conclude that the mean salary for female MBA graduates from this business school is lower than that for the male MBA graduates.

11.17 Let μ_1 denote the true mean approval rating for male players and μ_2 the true mean approval rating for female players.

$H_0: \mu_1 - \mu_2 = 0$ $H_a: \mu_1 - \mu_2 > 0$

$\alpha = 0.01$

Test statistic: $t = \dfrac{(\bar{x}_1 - \bar{x}_2) - 0}{\sqrt{\dfrac{s_1^2}{n_1} + \dfrac{s_2^2}{n_2}}}$

Assumptions: The sample sizes for each group is large (greater than or equal to 30) and the two samples are independently selected random samples.

$n_1 = 56$, $\bar{x}_1 = 2.76$, $s_1 = 0.44$, $n_2 = 67$, $\bar{x}_2 = 2.02$, $s_2 = 0.41$

$$t = \frac{(2.76 - 2.02) - 0}{\sqrt{\frac{(0.44)^2}{56} + \frac{(0.41)^2}{67}}} = \frac{0.74}{0.0772} = 9.58$$

$$df = \frac{\left(\frac{s_1^2}{n_1} + \frac{s_2^2}{n_2}\right)^2}{\frac{1}{n_1 - 1}\left(\frac{s_1^2}{n_1}\right)^2 + \frac{1}{n_2 - 1}\left(\frac{s_2^2}{n_2}\right)^2} = \frac{(0.003457 + 0.002509)^2}{\frac{(0.003457)^2}{55} + \frac{(0.002509)^2}{66}} = 113.8$$

So $df = 113$ (rounded down to an integer)

P-value = area under the 113 df t curve to the right of 9.58 $\approx 1 - 1 = 0$.

Since the P-value is less than α, the null hypothesis is rejected. At level of significance 0.05, the data supports the conclusion that the mean approval rating is higher for males than for females.

11.19   **a**   Let μ_1 denote the true mean hardness for chicken chilled 0 hours before cooking and μ_2 the true mean hardness for chicken chilled 2 hours before cooking.

H$_0$: $\mu_1 - \mu_2 = 0$ H$_a$: $\mu_1 - \mu_2 \neq 0$

$\alpha = 0.05$

Test statistic: $t = \dfrac{(\bar{x}_1 - \bar{x}_2) - 0}{\sqrt{\frac{s_1^2}{n_1} + \frac{s_2^2}{n_2}}}$

Assumptions: The sample sizes for each group is large (greater than or equal to 30) and the two samples are independently selected random samples.

$n_1 = 36$, $\bar{x}_1 = 7.52$, $s_1 = 0.96$, $n_2 = 36$, $\bar{x}_2 = 6.55$, $s_2 = 1.74$

$$t = \frac{(7.52 - 6.55) - 0}{\sqrt{\dfrac{(0.96)^2}{36} + \dfrac{(1.74)^2}{36}}} = \frac{0.97}{0.33121} = 2.93$$

$$df = \frac{\left(\dfrac{s_1^2}{n_1} + \dfrac{s_2^2}{n_2}\right)^2}{\dfrac{1}{n_1 - 1}\left(\dfrac{s_1^2}{n_1}\right)^2 + \dfrac{1}{n_2 - 1}\left(\dfrac{s_2^2}{n_2}\right)^2} = \frac{(0.0256 + 0.0841)^2}{\dfrac{(0.0256)^2}{35} + \dfrac{(0.0841)^2}{35}} = 54.5$$

So df = 54 (rounded down to an integer)

P-value = 2(area under the 54 df t curve to the right of 2.93) = 2(1 − 0.9975) = 2(0.00249) = 0.00498.

Since the P-value is less than α, the null hypothesis is rejected. At level of significance 0.05, there is sufficient evidence to conclude that there is a difference in mean hardness of chicken chilled 0 hours before cooking and chicken chilled 2 hours before cooking.

b   Let μ_1 denote the true mean hardness for chicken chilled 8 hours before cooking and μ_2 the true mean hardness for chicken chilled 24 hours before cooking.

$H_0: \mu_1 - \mu_2 = 0 \quad H_a: \mu_1 - \mu_2 \neq 0$

$\alpha = 0.05$

Test statistic: $t = \dfrac{(\bar{x}_1 - \bar{x}_2) - 0}{\sqrt{\dfrac{s_1^2}{n_1} + \dfrac{s_2^2}{n_2}}}$

Assumptions: The sample sizes for each group is large (greater than or equal to 30) and the two samples are independently selected random samples.

$n_1 = 36,\ \bar{x}_1 = 5.70,\ s_1 = 1.32,\ n_2 = 36,\ \bar{x}_2 = 5.65,\ s_2 = 1.50$

$$t = \frac{(5.70 - 5.65) - 0}{\sqrt{\dfrac{(1.32)^2}{36} + \dfrac{(1.50)^2}{36}}} = \frac{0.05}{0.333017} = 0.15$$

$$df = \dfrac{\left(\dfrac{s_1^2}{n_1} + \dfrac{s_2^2}{n_2}\right)^2}{\dfrac{1}{n_1-1}\left(\dfrac{s_1^2}{n_1}\right)^2 + \dfrac{1}{n_2-1}\left(\dfrac{s_2^2}{n_2}\right)^2} = \dfrac{(0.0484+0.0625)^2}{\dfrac{(0.0484)^2}{35} + \dfrac{(0.0625)^2}{35}} = 68.9$$

So $df = 68$ (rounded down to an integer)

P-value = 2(area under the 68 df t curve to the right of 0.15) = 2(1 − 0.5595)
= 2(0.44055) = 0.8811.

Since the P-value exceeds α, the null hypothesis is not rejected. At level of significance 0.05, there is not sufficient evidence to conclude that there is a difference in mean hardness of chicken chilled 8 hours before cooking and chicken chilled 24 hours before cooking.

c   Let μ_1 denote the true mean hardness for chicken chilled 2 hours before cooking and μ_2 the true mean hardness for chicken chilled 8 hours before cooking.

$n_1 = 36,\ \bar{x}_1 = 6.55,\ s_1 = 1.74,\ n_2 = 36,\ \bar{x}_2 = 5.70,\ s_2 = 1.32$

$$(6.55 - 5.70) \pm 1.669\sqrt{\dfrac{(1.74)^2}{36} + \dfrac{(1.32)^2}{36}}$$

$$\Rightarrow .85 \pm 1.669(.364005) \Rightarrow .85 \pm .6075 \Rightarrow (.242, 1.458)$$

Based on this sample, we believe that the mean hardness for chicken chilled for 2 hours before cooking is larger than the mean hardness for chicken chilled 8 hours before cooking. The difference may be as small as 0.242, or may be as large as 1.458.

11.21   Let μ_1 denote the true mean alkalinity for upstream locations and μ_2 the true mean alkalinity for downstream locations.

H_0: $\mu_1 - \mu_2 = -50$ H_a: $\mu_1 - \mu_2 < -50$

$\alpha = 0.05$

Test statistic: $t = \dfrac{(\bar{x}_1 - \bar{x}_2) - (-50)}{\sqrt{\dfrac{s_1^2}{n_1} + \dfrac{s_2^2}{n_2}}}$

Assumptions: The distribution of alkalinity is approximately normal for both types of sites (upstream and downstream) and the two samples are independently selected random samples.

$n_1 = 24$, $\overline{x}_1 = 75.9$, $s_1 = 1.83$, $n_2 = 24$, $\overline{x}_2 = 183.6$, $s_2 = 1.70$

$$t = \frac{(75.9 - 183.6) - (-50)}{\sqrt{\dfrac{(1.83)^2}{24} + \dfrac{(1.70)^2}{24}}} = \frac{-57.7}{0.50986} = 113.17$$

$$df = \frac{\left(\dfrac{s_1^2}{n_1} + \dfrac{s_2^2}{n_2}\right)^2}{\dfrac{1}{n_1 - 1}\left(\dfrac{s_1^2}{n_1}\right)^2 + \dfrac{1}{n_2 - 1}\left(\dfrac{s_2^2}{n_2}\right)^2} = \frac{(0.1395 + 0.1204)^2}{\dfrac{(0.1395)^2}{23} + \dfrac{(0.1204)^2}{23}} = 45.75$$

So $df = 45$ (rounded down to an integer)
P-value = area under the 45 df t curve to the right of $113 \approx 0$.

Since the P-value is less than α, the null hypothesis is rejected. The data supports the conclusion that the true mean alkalinity score for downstream sites is more than 50 units higher than that for upstream sites.

11.23   Let μ_1 denote the mean frequency of alcohol use for those that rush a sorority and μ_2 denote the mean frequency of alcohol use for those that do not rush a sorority.

H_0: $\mu_1 - \mu_2 = 0$ H_a: $\mu_1 - \mu_2 > 0$

$\alpha = 0.01$

Test statistic: $t = \dfrac{(\overline{x}_1 - \overline{x}_2) - 0}{\sqrt{\dfrac{s_1^2}{n_1} + \dfrac{s_2^2}{n_2}}}$

Assumptions: The sample size for each group is large (greater than or equal to 30) and the two samples are independently selected random samples.

$n_1 = 54$, $\overline{x}_1 = 2.72$, $s_1 = 0.86$, $n_2 = 51$, $\overline{x}_2 = 2.11$, $s_2 = 1.02$

$$t = \frac{(2.72 - 2.11) - 0}{\sqrt{\dfrac{(0.86)^2}{54} + \dfrac{(1.02)^2}{51}}} = \frac{0.61}{0.184652} = 3.30$$

$$df = \frac{\left(\dfrac{s_1^2}{n_1} + \dfrac{s_2^2}{n_2}\right)^2}{\dfrac{1}{n_1 - 1}\left(\dfrac{s_1^2}{n_1}\right)^2 + \dfrac{1}{n_2 - 1}\left(\dfrac{s_2^2}{n_2}\right)^2} = \frac{(0.0137 + 0.0204)^2}{\dfrac{(0.0137)^2}{53} + \dfrac{(0.0204)^2}{50}} = 98.002$$

So *df* = 98 (rounded down to an integer)

P-value = area under the 98 df t curve to the right of 3.30 = 1 − 0.9993 = 0.0007

Since the P-value is less than α, the null hypothesis is rejected. The data supports the conclusion that the true mean frequency of alcohol use is larger for those that rushed a sorority than for those who did not rush a sorority.

11.25 Let μ_1 denote the mean half-life of vitamin D in plasma for people on a normal diet. Let μ_2 denote the mean half-life of vitamin D in plasma for people on a high-fiber diet. Let $\mu_1 - \mu_2$ denote the true difference in mean half-life of vitamin D in plasma for people in these two groups (normal minus high fiber).

H₀: $\mu_1 - \mu_2 = 0$ Hₐ: $\mu_1 - \mu_2 > 0$

$\alpha = 0.01$

Test statistic: $t = \dfrac{(\bar{x}_1 - \bar{x}_2) - 0}{\sqrt{\dfrac{s_1^2}{n_1} + \dfrac{s_2^2}{n_2}}}$

Assumptions: The population distributions are (at least approximately) normal and the two samples are independently selected random samples.

Refer to the Minitab output given in the problem statement.

From the Minitab output the P-value = 0.007. Since the P-value is less than α, H₀ is rejected. There is sufficient evidence to conclude that the mean half-life of vitamin D is longer for those on a normal diet than for those on a high-fiber diet.

11.27   Let μ_1 denote the mean self-esteem score for students classified as having short duration loneliness. Let μ_2 denote the mean self-esteem score for students classified as having long duration loneliness.

$H_o: \mu_1 - \mu_2 = 0$ $H_a: \mu_1 - \mu_2 > 0$

$\alpha = 0.01$

Test statistic: $t = \dfrac{(\bar{x}_1 - \bar{x}_2) - 0}{\sqrt{\dfrac{s_1^2}{n_1} + \dfrac{s_2^2}{n_2}}}$

Assumptions: The population distributions are (at least approximately) normal and the two samples are independently selected random samples.

$n_1 = 72,\ \bar{x}_1 = 76.78\ ,\ s_1 = 17.8,\ n_2 = 17,\ \bar{x}_2 = 64.00,\ s_2 = 15.68$

$$t = \frac{(76.78 - 64.00) - 0}{\sqrt{\dfrac{(17.8)^2}{72} + \dfrac{(15.68)^2}{17}}} = \frac{12.78}{4.34316} = 2.9426$$

$$df = \frac{\left(\dfrac{s_1^2}{n_1} + \dfrac{s_2^2}{n_2}\right)^2}{\dfrac{1}{n_1 - 1}\left(\dfrac{s_1^2}{n_1}\right)^2 + \dfrac{1}{n_2 - 1}\left(\dfrac{s_2^2}{n_2}\right)^2} = \frac{(4.4006 + 14.4625)^2}{\dfrac{(4.4006)^2}{71} + \dfrac{(14.4625)^2}{16}} = 26.7$$

So $df = 26$ (rounded down to an integer)

P-value = area under the 26 df t curve to the right of $2.9426 \approx 0.0034$.

Since the P-value is less than α, H_o is rejected. The sample data supports the conclusion that the mean self esteem is lower for students classified as having long duration loneliness than for students classified as having short duration loneliness.

Exercises 11.29 – 11.40

11.29   a   If possible, treat each patient with both drugs with one drug used on one eye and the other drug used on the other eye. Then take observations (readings) of eye pressure on each eye. If this treatment method is not possible, then request the ophthalmologist to pair patients according to their eye pressure so that the two people in a pair have approximately equal eye pressure. Then treat one of the patients in the pair with the new drug and record the reduction in eye pressure. Treat the other person in that pair with the standard treatment and record the reduction in eye pressure. These two readings would constitute a pair. Repeat for each of the other pairs to obtain the paired sample data.

b   Both procedures above would result in paired data.

c   Select a group of persons to participate in the study. Randomly select a subset of this group to receive the new drug, and give the standard treatment to the remaining people. Measure reduction in eye pressure for both groups. The resulting observations would constitute independent samples.

This experiment is probably not as informative as a paired experiment when the same total number of data values are collected due to patient to patient variability which can be quite large.

11.31 a   Let d = the difference in the percent of exams earning college credit at each central coach high school i.e. % in 1997 - % in 2002.

Let μ_d denote the mean difference in the percent of exams earning college credit at each central coach high school i.e. % in 1997 - % in 2002

$H_0: \mu_d = 0$ $H_a: \mu_d > 0$

$\alpha = 0.05$ (not stated in the question, but used for demonstration purposes)

Assumptions: The sample is random and independent. The sample size is small, but a boxplot of the differences shows symmetry, suggesting that distribution of the population differences are approximately normal.

The test statistic is: $t = \dfrac{\bar{x}_d - 0}{\dfrac{s_d}{\sqrt{n}}}$ with d.f. = 6

$\bar{d} = $ -5.4 and $s_d = 12.05$

$t = \dfrac{-5.4 - 0}{\dfrac{12.05}{\sqrt{7}}} = -1.15$

P-value = the area under the 6 df t curve to the right of -1.15) $\approx$ 0.853.

Thus, the null hypothesis cannot be rejected at 0.05 There is not sufficient evidence to suggest that the mean difference in the percent of exams earning college credit at each central coach high school has declined between 1997 and 2002.

b   No. These 7 schools are representative of high schools located on the central coast of California, but they are not representative of all California high schools.

c   After computing the differences, there is clearly an outlier (school #5). With such a small sample size, this violates one of the assumptions of the t-test. It would be doubtful that the sample of differences would come from an approximately normal population and therefore a t-test would not be appropriate.

11.33 Let μ_d denote the mean difference in verbal ability in children born prematurely (aged 8 – aged 3)

H$_o$: $\mu_d = 0$ H$_a$: $\mu_d > 0$
$\alpha = 0.05$ (not stated, but used for demonstration)

From MINITAB: t = 3.17, P-value = 0.001

Thus, the null hypothesis is rejected. There is sufficient evidence to suggest that the mean verbal ability in children born prematurely increases between the ages of 3 and 8.

11.35 **a**   The data are paired because the response for the number of science courses each girl in the sample intended to take is logically matched with the same girl's response for the number of science courses she thought boys should take.

b   Let μ$_d$ denote the true average difference in the intended number of courses for girls and boys (girls – boys).

The 95% confidence interval for μ$_d$ is

$$\bar{d} \pm (t \text{ critical})\frac{S_d}{\sqrt{n}} \Rightarrow -0.83 \pm (1.971)\left(\frac{1.51}{\sqrt{223}}\right)$$

$$\Rightarrow -0.83 \pm 0.1988 \Rightarrow (-1.029, -0.631).$$

With 95% confidence, it is estimated that the mean difference in the number of science courses girls intend to take and what they think boys should take is –1.029 and –0.631.

11.37 **a** Even though the data are paired the two responses from the same subject, one in 1994 and one in 1995, are most likely not strongly "correlated". The questionnaire asked about alcohol consumption during the "previous week" but the alcohol consumption pattern may vary quite a bit within the same individual from one week to the next which would explain the low correlation between the paired responses.

b Let μ_d denote the true average difference in the number of drinks consumed by this population between 1994 and 1995 (average for 1994 – average for 1995).

A 95% confidence interval for μ_d is

$$\bar{d} \pm (t \text{ critical})\frac{s_d}{\sqrt{n}} \Rightarrow 0.38 \pm (1.985)\left(\frac{5.52}{\sqrt{96}}\right)$$

$$\Rightarrow 0.38 \pm 1.1185 \Rightarrow (-0.738, 1.498).$$

Since zero is included in the confidence interval, zero is a plausible value for μ_d and hence the data do not provide evidence indicating a decrease in the mean number of drinks consumed.

c Let μ_d denote the mean difference in number of drinks consumed by non credit card shoppers between 1994 and 1995.

$H_0: \mu_d = 0$ $H_a: \mu_d \neq 0$

We will compute a P-value for this test.

The test statistic is: $t = \dfrac{\bar{x}_d - 0}{\dfrac{s_d}{\sqrt{n}}}$ with d.f. = 849

$\bar{d} = 0.12$ and $s_d = 4.58$

$$t = \frac{0.12 - 0}{\dfrac{4.58}{\sqrt{850}}} = 0.764$$

P-value =2 times (the area under the 849 df t curve to the right of 0.764) $\approx$ 0.445.

Thus, the null hypothesis cannot be rejected at any of the commonly used significance levels (e.g., α = 0.01, 0.05, or 0.10). There is not sufficient evidence to support the conclusion that the mean number of drinks consumed by the non credit card shoppers has changed between 1994 and 1995.

11.39 Let μ_d denote the true average difference in number of seeds detected by the two methods (Direct – Stratified).

H_0: $\mu_d = 0$ (no difference in average number of seeds detected)

H_a: $\mu_d \neq 0$ (average number of seeds detected by the Direct method is not the same as the average number of seeds detected by the Stratified method)

$\alpha = 0.05$

The test statistic is: $t = \dfrac{\overline{x}_d - 0}{\dfrac{s_d}{\sqrt{n}}}$ with d.f. = 26

The differences are: 16, –4, –8, 4, –32, 0, 12, 0, 4, –8, 4, 12, 8, –28, 4, 0, 0, 4, 0, –8, –8, 0, 0, –4, –28, 4, –36.

From these: $\overline{x}_d = -3.407$ and $s_d = 13.253$

$$t = \frac{-3.407 - 0}{\dfrac{13.253}{\sqrt{27}}} = -1.34$$

P-value = 2(area under the 26 df t curve to the left of –1.34) $\approx$ 2(0.096) = 0.192.

Since the P-value exceeds α, the null hypothesis is not rejected. The data do not provide sufficient evidence to conclude that the mean number of seeds detected differs for the two methods.

Exercises 11.41 – 11.56

11.41 Let π_1 denote the proportion of students who registered by phone that were satisfied with the registration process and π_2 denote the corresponding proportion for those who registered on-line.

H_0: $\pi_1 - \pi_2 = 0$ H_a: $\pi_1 - \pi_2 < 0$

$\alpha = 0.05$

$$z = \frac{p_1 - p_2}{\sqrt{\frac{p_c(1-p_c)}{n_1} + \frac{p_c(1-p_c)}{n_2}}}$$

$$p_1 = \frac{57}{80} = 0.7125 \qquad p_2 = \frac{50}{60} = 0.8333$$

$$p_c = \frac{n_1 p_1 + n_2 p_2}{n_1 + n_2} = \frac{57 + 50}{80 + 60} = 0.7643$$

$$z = \frac{(0.7125 - 0.8333)}{\sqrt{\frac{0.7643(1-0.7643)}{80} + \frac{0.7643(1-0.7643)}{60}}} = \frac{-0.1208}{0.0725} = -1.666$$

P-value = Area under the z curve to the left of −1.666 = 0.0479.

Since the P-value is less than α, H_o is rejected. The data supports the claim that the proportion of satisfied students is higher for those who registered on-line than for those who registered over the phone.

11.43 Let π_1 denote the proportion of resumes with white sounding names receiving positive responses, and π_2 denote the proportion of resumes with black sounding names receiving positive responses.

$H_o: \pi_1 - \pi_2 = 0 \qquad H_a: \pi_1 - \pi_2 > 0$

$\alpha = 0.05$ (for demonstration purposes)

$$z = \frac{p_1 - p_2}{\sqrt{\frac{p_c(1-p_c)}{n_1} + \frac{p_c(1-p_c)}{n_2}}}$$

$$p_1 = \frac{250}{2500} = 0.1 \qquad p_2 = \frac{167}{2500} = 0.0668$$

$$p_c = \frac{n_1 p_1 + n_2 p_2}{n_1 + n_2} = \frac{250 + 167}{5000} = 0.0844$$

$$z = \frac{(0.1 - 0.0668)}{\sqrt{\frac{0.0834(1-0.0834)}{2500} + \frac{0.0834(1-0.0834)}{2500}}} = \frac{0.0332}{0.0078} = 4.25$$

P-value = Area under the z curve to the right of 4.25 ≈ 0

Since the P-value is less than α, H₀ is rejected. There is enough evidence to suggest that the proportion receiving positive responses is higher for those resumes with white-sounding first names.

11.45   **a**   Let π_1 denote the true proportion of all high risk patients who receive insulin and develop diabetes and π_2 denote the true proportion of all high risk patients who receive do not receive insulin and develop diabetes.

$$n_1 = 169, \; x_1 = 25, \; p_1 = \frac{25}{169} = 0.1479, \; n_2 = 170, \; x_2 = 24, \; p_2 = \frac{24}{170} = 0.1412$$

The 90% confidence interval for $\pi_1 - \pi_2$ is

$$(0.1479 - 0.1412) \pm 1.645 \sqrt{\frac{0.1479(0.8521)}{169} + \frac{0.1412(0.8588)}{170}} \Rightarrow 0.0067 \pm 1.645(0.0382)$$
$$\Rightarrow 0.0067 \pm 0.0628 \Rightarrow (-0.0561, 0.0695).$$

b   With 90% confidence, it is estimated that the true proportion of patients developing diabetes may be as much as 0.0695 more in the insulin group than in the control group; but it also may be as much as 0.0561 less in the insulin group as in the control group.

c   Because 0 is in the interval, it is possible that there is no difference in the proportion of the patients developing diabetes in the two groups. The proposed treatment doesn't appear very effective.

11.47   Let π_1 denote the proportion of elementary school teachers who are very satisfied and π_2 denote the proportion of high school teachers who are very satisfied.

$H_0: \pi_1 - \pi_2 = 0$ $H_a: \pi_1 - \pi_2 \neq 0$

$\alpha = 0.05$

$$z = \frac{p_1 - p_2}{\sqrt{\dfrac{p_c(1-p_c)}{n_1} + \dfrac{p_c(1-p_c)}{n_2}}}$$

$$p_1 = \frac{224}{395} = 0.567089 \qquad p_2 = \frac{126}{266} = 0.473684$$

$$p_c = \frac{n_1 p_1 + n_2 p_2}{n_1 + n_2} = \frac{224 + 126}{395 + 266} = 0.529501$$

$$z = \frac{(0.567089 - 0.473684)}{\sqrt{\dfrac{0.529501(0.470499)}{395} + \dfrac{0.529501(0.470499)}{266}}} = \frac{0.093404}{0.039589} = 2.36$$

P-value = 2(area under the z curve to the right of 2.36) = 2(1 – 0.9909) = 0.0182.

Since the P-value is less than α, H₀ is rejected. The data supports the claim that the proportion of teachers who are "very satisfied" is different for elementary-school teachers than for high-school teachers.

11.49 The decision was based on all an analysis of all the soldiers that went served in the Gulf War (the population). Because a census was performed, no inference procedure was necessary.

11.51 Let π_1 denote the proportion of females who are concerned about getting AIDS and let π_2 denote the proportion of males who are similarly concerned.

H₀: $\pi_1 - \pi_2 = 0$ Hₐ: $\pi_1 - \pi_2 > 0$

We will compute P-value for this test.

$$z = \frac{p_1 - p_2}{\sqrt{\dfrac{p_c(1 - p_c)}{n_1} + \dfrac{p_c(1 - p_c)}{n_2}}}$$

$p_1 = 0.427$, $p_2 = 0.275$,

$$p_c = \frac{n_1 p_1 + n_2 p_2}{n_1 + n_2} = \frac{568(0.427) + 234(0.275)}{568 + 234} = 0.3827$$

$$z = \frac{(0.427 - 0.275)}{\sqrt{\dfrac{0.3827(1 - 0.3827)}{568} + \dfrac{0.3827(1 - 0.3827)}{234}}} = \frac{0.152}{0.0378} = 4.026$$

P-value = area under the z curve to the right of 4.026 ≈ 0.00003.

216

Since the P-value is much smaller than any of the commonly used significance values, H_o is rejected. There is sufficient evidence in the data to support the conclusion that the proportion of females who are concerned about getting AIDS is greater than the proportion of males so concerned.

11.53 Let π_1 denote the proportion of students in the College of Computing who lose their HOPE scholarship at the end of the first year and let π_2 denote the proportion of students in the Ivan Allen College who lose their HOPE scholarship at the end of the first year.

H_o: $\pi_1 - \pi_2 = 0$ H_a: $\pi_1 - \pi_2 \neq 0$

We will compute a P-value for this test.

$$z = \frac{p_1 - p_2}{\sqrt{\dfrac{p_c(1-p_c)}{n_1} + \dfrac{p_c(1-p_c)}{n_2}}}$$

$p_1 = 0.532$, $p_2 = 0.649$,

$$p_c = \frac{n_1 p_1 + n_2 p_2}{n_1 + n_2} = \frac{137(0.532) + 111(0.649)}{137 + 111} = 0.5842$$

$$z = \frac{(0.532 - 0.649)}{\sqrt{\dfrac{0.5842(1-0.5842)}{137} + \dfrac{0.5842(1-0.5842)}{111}}} = \frac{-0.11665}{0.06294} = -1.853$$

P-value = 2 times (the area under the z curve to the left of −1.853) ≈ 0.0638.

H_o cannot be rejected at a significance level of 0.05 or smaller. There is not sufficient evidence in the data to support the conclusion that the proportion of students in the College of Computing who lose their HOPE scholarship at the end of one year is different from the proportion for the Ivan Allen College.

11.55 Let π_1 denote the true proportion of returning students who do not take an orientation course and π_2 denote the true proportion of returning students who do take an orientation course.

$$n_1 = 94, \; x_1 = 50, \; p_1 = \frac{50}{94} = 0.5319, \; n_2 = 94, \; x_2 = 56, \; p_2 = \frac{56}{94} = 0.5957$$

The 95% confidence interval for $\pi_1 - \pi_2$ is

$$(0.5319 - 0.5957) \pm 1.96 \sqrt{\frac{0.5319(0.4681)}{94} + \frac{0.5957(0.4043)}{94}} \Rightarrow -0.0638 \pm 1.96(0.0722)$$

$$\Rightarrow -0.0638 \pm 0.1415 \Rightarrow (-0.2053, 0.0777).$$

With 95% confidence, it is estimated that the difference between the proportion of returning who do not take an orientation course and the proportion of returning students who do take an orientation course may be as small as –0.2053 to as large as 0.0777.

Exercises 11.57 – 11.63

11.57 Let μ_1 denote the true average fluoride concentration for livestock grazing in the polluted region and μ_2 denote the true average fluoride concentration for livestock grazing in the unpolluted regions.

H_0: $\mu_1 - \mu_2 = 0$ H_a: $\mu_1 - \mu_2 > 0$

$\alpha = 0.05$

The test statistic is: rank sum for polluted area (sample 1).

Sample	Ordered Data	Rank
2	14.2	1
1	16.8	2
1	17.1	3
2	17.2	4
2	18.3	5
2	18.4	6
1	18.7	7
1	19.7	8
2	20.0	9
1	20.9	10
1	21.3	11
1	23.0	12

Rank sum = $(2 + 3 + 7 + 8 + 10 + 11 + 12) = 53$

P-value: This is an upper-tail test. With $n_1 = 7$ and $n_2 = 5$, Appendix Table VI tells us that the P-value > 0.05.

Since the P-value exceeds α, H_o is not rejected. The data does not support the conclusion that there is a larger average fluoride concentration for the polluted area than for the unpolluted area.

11.59 **a** Let μ_1 denote the true average ascent time using the lateral gait and μ_2 denote the true average ascent time using the four-beat diagonal gait.

H_o: $\mu_1 - \mu_2 = 0$ H_a: $\mu_1 - \mu_2 \neq 0$

A value for α was not specified in the problem, so a value of 0.05 was chosen for illustration.

The test statistic is: Rank sum for diagonal gait.

Gait	Ordered Data	Rank
D	0.85	1
L	0.86	2
L	1.09	3
D	1.24	4
D	1.27	5
L	1.31	6
L	1.39	7
D	1.45	8
L	1.51	9
L	1.53	10
L	1.64	11
D	1.66	12
D	1.82	13

Rank sum = $1 + 4 + 5 + 8 + 12 + 13 = 43$

P-value: This is a two-tail test. With $n_1 = 7$ and $n_2 = 6$, Appendix Table VI tells us that the P-value > 0.05.

Since the P-value exceeds α, H_o is not rejected. The data does not suggest that there is a difference in mean ascent time for the diagonal and lateral gaits.

  b We can be at least 95% confident (actually 96.2% confident) that the difference in the mean ascent time using lateral gait and the mean ascent time using diagonal gait may be as small as −.43 to as large as 0.3697.

11.61 Let μ_1 denote the true mean number of binges per week for people who use Imipramine and μ_2 the true mean number of binges per week for people who use a placebo.

H_0: $\mu_1 - \mu_2 = 0$ H_a: $\mu_1 - \mu_2 < 0$

$\alpha = 0.05$

The test statistic is: Rank sum for the Imipramine group.

Group	Ordered Data	Rank
I	1	1.5
I	1	1.5
I	2	3.5
I	2	3.5
I	3	6
P	3	6
P	3	6
P	4	8.5
P	4	8.5
I	5	10
P	6	11
I	7	12
P	8	13
P	10	14
I	12	15
P	15	16

Rank sum = 1.5 + 1.5 + 3.5 + 3.5 + 6 + 10 + 12 + 15 = 53

P-value: This is an lower-tail test. With $n_1 = 8$ and $n_2 = 8$, Appendix Table VI tells us that the P-value > 0.05.

Since the P-value exceeds α, H_0 is not rejected. The data does not provide enough evidence to suggest that Imipramine is effective in reducing the mean number of binges per week.

Exercises 11.64 – 11.80

11.65   Let π_1 denote the true proportion of children drinking fluoridated water who have decayed teeth, and let π_2 denote the true proportion of children drinking non-fluoridated water who have decayed teeth.

$n_1 = 119$ $x_1 = 67$ $p_1 = \dfrac{67}{119} = 0.5630$ $n_2 = 143$ $x_2 = 106$ $p_2 = \dfrac{106}{143} = 0.7413$

The 90% confidence interval for $\pi_1 - \pi_2$ is

$$(0.5630 - 0.7413) \pm 1.645 \sqrt{\frac{0.5630(0.4370)}{119} + \frac{0.7413(0.2587)}{143}} \Rightarrow -0.1783 \pm 1.645(0.0584)$$

$$\Rightarrow -0.1783 \pm 0.096 \Rightarrow (-0.2743, -0.0823).$$

The interval does not contain 0, so we can conclude that the two true proportions differ. Since both endpoints of the interval are negative, this indicates that $\pi_1 < \pi_2$. Thus with 90% confidence, it is estimated that the percentage of children drinking fluoridated water that have decayed teeth is less than that for children drinking non-fluoridated water by as little as about 8%, to as much as 27%.

11.67 **a** Let μ_1 denote the true average peak loudness for open-mouth chewing and μ_2 the true average peak loudness for closed mouth chewing. Then $\mu_1 - \mu_2$ denotes the difference between the means of open-mouthed and closed-mouth chewing.

$n_1 = 10$, $\bar{x}_1 = 63$, $s_1 = 13$, $n_2 = 10$, $\bar{x}_2 = 54$, $s_2 = 16$

$$V_1 = \frac{s_1^2}{n_1} = \frac{(13)^2}{10} = 16.9 \quad V_2 = \frac{s_2^2}{n_2} = \frac{(16)^2}{10} = 25.6$$

$$df = \frac{(V_1 + V_2)^2}{\frac{V_1^2}{n_1 - 1} + \frac{V_2^2}{n_2 - 1}} = \frac{(16.9 + 25.6)^2}{\frac{(16.9)^2}{9} + \frac{(25.6)^2}{9}} = \frac{1806.25}{104.552222} = 17.276$$

Use df = 17.

The 95% confidence interval for $\mu_1 - \mu_2$ based on this sample is

$$(63 - 54) \pm 2.11\sqrt{16.9 + 25.6} \Rightarrow 9 \pm 2.11(6.519202) \Rightarrow 9 \pm 13.75 \Rightarrow (-4.75, 22.75).$$

Observe that the interval includes 0, and so 0 is one of the plausible values of $\mu_1 - \mu_2$. That is, it is plausible that there is no difference in the mean loudness for open-mouth and closed-mouth chewing of potato chips.

b Let μ_1 denote the true average peak loudness for closed-mouth chewing of potato chips and μ_2 the true average peak loudness for closed-mouth chewing of tortilla chips.

$H_0: \mu_1 - \mu_2 = 0 \quad H_a: \mu_1 - \mu_2 \neq 0$

221

$\alpha = 0.01$

Test statistic: $t = \dfrac{(\bar{x}_1 - \bar{x}_2) - 0}{\sqrt{\dfrac{s_1^2}{n_1} + \dfrac{s_2^2}{n_2}}}$

Assumptions: The population distributions are (at least approximately) normal and the two samples are independently selected random samples.

$n_1 = 10,\ \bar{x}_1 = 54,\ s_1 = 16,\ n_2 = 10,\ \bar{x}_2 = 53,\ s_2 = 16$

$t = \dfrac{(54 - 53) - 0}{\sqrt{\dfrac{(16)^2}{10} + \dfrac{(16)^2}{10}}} = \dfrac{1.0}{7.1554} = 0.1398$

$df = \dfrac{\left(\dfrac{s_1^2}{n_1} + \dfrac{s_2^2}{n_2}\right)^2}{\dfrac{1}{n_1 - 1}\left(\dfrac{s_1^2}{n_1}\right)^2 + \dfrac{1}{n_2 - 1}\left(\dfrac{s_2^2}{n_2}\right)^2} = \dfrac{(25.6 + 25.6)^2}{\dfrac{(25.6)^2}{9} + \dfrac{(25.6)^2}{9}} = 18$

So $df = 18$ (rounded down to an integer)

P-value = 2(area under the 18 df t curve to the right of 0.1398) $\approx$ 2(0.445) = 0.890.

Since the P-value exceeds α, H$_0$ is not rejected. There is not sufficient evidence to conclude that there is a difference in the mean peak loudness for closed-mouth chewing of tortilla chips and potato chips.

c Let μ_1 denote the true average peak loudness for fresh tortilla chips when chewing closed-mouth. Let μ_2 denote the true average peak loudness of stale tortilla chips when chewing closed-mouth.

H$_0$: $\mu_1 - \mu_2 = 0$ H$_a$: $\mu_1 - \mu_2 > 0$

$\alpha = 0.05$

Test statistic: $t = \dfrac{(\bar{x}_1 - \bar{x}_2) - 0}{\sqrt{\dfrac{s_1^2}{n_1} + \dfrac{s_2^2}{n_2}}}$

Assumptions: The population distributions are (at least approximately) normal and the two samples are independently selected random samples.

$n_1 = 10$, $\bar{x}_1 = 56$, $s_1 = 14$, $n_2 = 10$, $\bar{x}_2 = 53$, $s_2 = 16$

$$t = \frac{(56-53)-0}{\sqrt{\dfrac{(14)^2}{10}+\dfrac{(16)^2}{10}}} = \frac{3.0}{6.723} = 0.4462$$

$$df = \frac{\left(\dfrac{s_1^2}{n_1}+\dfrac{s_2^2}{n_2}\right)^2}{\dfrac{1}{n_1-1}\left(\dfrac{s_1^2}{n_1}\right)^2+\dfrac{1}{n_2-1}\left(\dfrac{s_2^2}{n_2}\right)^2} = \frac{(19.6+25.6)^2}{\dfrac{(19.6)^2}{9}+\dfrac{(25.6)^2}{9}} = 17.69$$

So $df = 17$ (rounded down to an integer)

P-value = area under the 17 df t curve to the right of $0.4462 \approx 0.331$.

Since the P-value exceeds α, H_0 is not rejected. There is not sufficient evidence to conclude that there is a difference in the mean peak loudness when chewing fresh or stale tortilla chips closed-mouth.

11.69 Let μ_1 denote the mean number of imitations for infants who watch a human model. Let μ_2 denote the mean number of imitations for infants who watch a doll.

$H_0: \mu_1 - \mu_2 = 0$ $H_a: \mu_1 - \mu_2 > 0$
$\alpha = 0.01$

Test statistic: $t = \dfrac{(\bar{x}_1 - \bar{x}_2)-0}{\sqrt{\dfrac{s_1^2}{n_1}+\dfrac{s_2^2}{n_2}}}$

Assumptions: The population distributions are (at least approximately) normal and the two samples are independently selected random samples.

$n_1 = 12, \bar{x}_1 = 5.14, s_1 = 1.6, n_2 = 15, \bar{x}_2 = 3.46, s_2 = 1.3$

$$t = \frac{(5.14-3.46)-0}{\sqrt{\frac{(1.6)^2}{12}+\frac{(1.3)^2}{15}}} = \frac{1.68}{0.570964} = 2.94$$

$$df = \frac{\left(\frac{s_1^2}{n_1}+\frac{s_2^2}{n_2}\right)^2}{\frac{1}{n_1-1}\left(\frac{s_1^2}{n_1}\right)^2 + \frac{1}{n_2-1}\left(\frac{s_2^2}{n_2}\right)^2} = \frac{(0.1024+0.0676)^2}{\frac{(0.1024)^2}{11}+\frac{(0.0676)^2}{14}} = 21.1$$

So df = 21 (rounded down to an integer)

P-value = area under the 21 df t curve to the right of 2.94 ≈ 0.0039.

The P-value is less than 0.01 and so the null hypothesis is rejected. The data supports the conclusion that the mean number of imitations by infants who watch a human model is larger than the mean number of imitations by infants who watch a doll.

11.71   **a** $t = \frac{(8.63-10.97)-0}{\sqrt{\frac{(7.1)^2}{30}+\frac{(8.9)^2}{30}}} = \frac{-2.34}{2.0786} = -1.126$, which rounds to t = −1.13.

     b   Let μ_1 be the mean EAT score for models and μ_2 the mean EAT score for controls.

       H$_o$: $\mu_1 - \mu_2 = 0$ H$_a$: $\mu_1 - \mu_2 < 0$

       α = 0.01 (A value for α is not specified in the problem. We use α=0.01 for illustration)

       Test statistic: $t = \frac{(\bar{x}_1 - \bar{x}_2)-0}{\sqrt{\frac{s_1^2}{n_1}+\frac{s_2^2}{n_2}}}$

       Assumptions: The population distributions are (at least approximately) normal and the two samples are independently selected random samples.

       n_1 = 30, $\bar{x}_1$ = 8.63, s_1 = 7.1, n_1 = 30, $\bar{x}_2$ = 10.97, s_2 = 8.9

$$t = \frac{(8.63 - 10.97) - 0}{\sqrt{\frac{(7.1)^2}{30} + \frac{(8.9)^2}{30}}} = \frac{-2.34}{2.0786} = -1.126$$

$$df = \frac{\left(\frac{s_1^2}{n_1} + \frac{s_2^2}{n_2}\right)^2}{\frac{1}{n_1 - 1}\left(\frac{s_1^2}{n_1}\right)^2 + \frac{1}{n_2 - 1}\left(\frac{s_2^2}{n_2}\right)^2} = \frac{(1.68033 + 2.64033)^2}{\frac{(1.68033)^2}{29} + \frac{(2.64033)^2}{29}} = 55.27$$

So $df = 55$ (rounded down to an integer)

P-value = area under the 55 df t curve to the left of $-1.13 = 0.13$. Since the P-value exceeds α, H_0 is not rejected. The data does not support the conclusion that models have a smaller mean EAT value than do non-models.

c H_0: $\mu_1 - \mu_2 = 0$ H_a: $\mu_1 - \mu_2 > 0$

$$t = \frac{(10.97 - 8.63) - 0}{\sqrt{\frac{(8.9)^2}{30} + \frac{(7.1)^2}{30}}} = \frac{2.34}{2.0786} = 1.13$$

df will still be equal to 55.

P-value = area under the 55 df t curve to the right of $1.13 = 1 - 0.87 = 0.13$. Hence the value of t now is the negative of the value of t in part **a**, but the P-value is the same as that in part **b**.

11.73   **a**   Let μ_1 denote the true mean self-esteem score for students hired by the university as RA's and let μ_2 denote the true mean self-esteem score for those not hired as RA's.

H_0: $\mu_1 - \mu_2 = 0$ H_a: $\mu_1 - \mu_2 \neq 0$

$\alpha = 0.05$

Test statistic: $t = \dfrac{(\bar{x}_1 - \bar{x}_2) - 0}{\sqrt{\dfrac{s_1^2}{n_1} + \dfrac{s_2^2}{n_2}}}$

Assumptions: The sample sizes for each group is large (greater than or equal to 30) and the two samples are independently selected random samples.

$n_1 = 69$, $\bar{x}_1 = 83.28$, $s_1 = 12.21$, $n_2 = 47$, $\bar{x}_2 = 81.96$, $s_2 = 12.78$

$$t = \frac{(83.28 - 81.96) - 0}{\sqrt{\dfrac{(12.21)^2}{69} + \dfrac{(12.78)^2}{47}}} = \frac{1.32}{2.373965} = 0.56$$

$$df = \frac{\left(\dfrac{s_1^2}{n_1} + \dfrac{s_2^2}{n_2}\right)^2}{\dfrac{1}{n_1 - 1}\left(\dfrac{s_1^2}{n_1}\right)^2 + \dfrac{1}{n_2 - 1}\left(\dfrac{s_2^2}{n_2}\right)^2} = \frac{(2.1606 + 3.4751)^2}{\dfrac{(2.1606)^2}{68} + \dfrac{(3.4751)^2}{46}} = 95.90$$

So $df = 95$ (rounded down to an integer)

P-value = 2(area under the 95 df t curve to the right of 0.56) = 2(0.28975) = 0.5795.

Since the P-value exceeds α, the null hypothesis is not rejected. The sample data does not support the conclusion that the mean self-esteem score for students hired as RA's differs from that of students not hired as RA's.

b   Let μ_1 denote the true mean leadership score for students hired by the university as RA's and let μ_2 denote the true mean leadership score for those not hired as RA's.

H_0: $\mu_1 - \mu_2 = 0$ H_a: $\mu_1 - \mu_2 \neq 0$

$\alpha = 0.05$

Test statistic: $t = \dfrac{(\bar{x}_1 - \bar{x}_2) - 0}{\sqrt{\dfrac{s_1^2}{n_1} + \dfrac{s_2^2}{n_2}}}$

Assumptions: The sample sizes for each group is large (greater than or equal to 30) and the two samples are independently selected random samples.

$n_1 = 69$, $\bar{x}_1 = 62.51$, $s_1 = 3.05$, $n_2 = 47$, $\bar{x}_2 = 62.43$, $s_2 = 3.36$

$$t = \frac{(62.51 - 62.43) - 0}{\sqrt{\dfrac{(3.05)^2}{69} + \dfrac{(3.36)^2}{47}}} = \frac{0.08}{0.612391} = 0.1306$$

$$df = \frac{\left(\dfrac{s_1^2}{n_1} + \dfrac{s_2^2}{n_2}\right)^2}{\dfrac{1}{n_1-1}\left(\dfrac{s_1^2}{n_1}\right)^2 + \dfrac{1}{n_2-1}\left(\dfrac{s_2^2}{n_2}\right)^2} = \frac{(0.13482 + 0.24020)^2}{\dfrac{(0.13482)^2}{68} + \dfrac{(0.24020)^2}{46}} = 92.4$$

So df = 92 (rounded down to an integer)

P-value = 2(area under the 92 df t curve to the right of 0.1306) = 2(0.4482) = 0.8964.

Since the P-value exceeds α, the null hypothesis is not rejected. The sample data does not support the conclusion that the mean leadership score for students hired as RA's differs from that of students not hired as RA's.

c   Let μ_1 denote the true mean GPA for students hired by the university as RA's and μ_2 the true mean GPA for those not hired as RA's.

H$_0$: $\mu_1 - \mu_2 = 0$ H$_a$: $\mu_1 - \mu_2 \neq 0$

$\alpha = 0.05$

Test statistic: $t = \dfrac{(\bar{x}_1 - \bar{x}_2) - 0}{\sqrt{\dfrac{s_1^2}{n_1} + \dfrac{s_2^2}{n_2}}}$

Assumptions: The sample sizes for each group is large (greater than or equal to 30) and the two samples are independently selected random samples.

n_1 = 69, $\bar{x}_1$ = 2.94, s_1 = 0.61, n_2 = 47, $\bar{x}_2$ = 2.60, s_2 = 0.79

$$t = \frac{(2.94 - 2.60) - 0}{\sqrt{\dfrac{(0.61)^2}{69} + \dfrac{(0.79)^2}{47}}} = \frac{0.34}{0.136644} = 2.4882$$

$$df = \frac{\left(\dfrac{s_1^2}{n_1} + \dfrac{s_2^2}{n_2}\right)^2}{\dfrac{1}{n_1-1}\left(\dfrac{s_1^2}{n_1}\right)^2 + \dfrac{1}{n_2-1}\left(\dfrac{s_2^2}{n_2}\right)^2} = \frac{(0.00539 + 0.01328)^2}{\dfrac{(0.00539)^2}{68} + \dfrac{(0.01328)^2}{46}} = 81.8$$

So df = 81 (rounded down to an integer)

P-value = 2(area under the 81 df t curve to the right of 2.4882) = 2(0.0074) = 0.0148.

Since the P-value is less than α, the null hypothesis is rejected. The sample data supports the conclusion that the mean GPA for students hired as RA's differs from that of students not hired as RA's.

11.75 **a** Let π_1 denote the true proportion of undergraduate males who perceive pressure for a date as sexual harassment and π_2 denote the true proportion of undergraduate females who perceive pressure for a date as sexual harassment.

$H_0: \pi_1 - \pi_2 = 0$ $H_a: \pi_1 - \pi_2 < 0$

$\alpha = 0.05$

$$z = \frac{p_1 - p_2}{\sqrt{\dfrac{p_c(1-p_c)}{n_1} + \dfrac{p_c(1-p_c)}{n_2}}}$$

$n_1 = 1336$, $x_1 = 882$, $p_1 = \dfrac{882}{1336} = 0.6602$, $n_2 = 1346$, $x_2 = 1104$, $p_2 = \dfrac{1104}{1346} = 0.8202$

$$p_c = \frac{n_1 p_1 + n_2 p_2}{n_1 + n_2} = \frac{882 + 1104}{1336 + 1346} = 0.7405$$

$$z = \frac{(0.6602 - 0.8202)}{\sqrt{\dfrac{0.7405(0.2595)}{1336} + \dfrac{0.7405(0.2595)}{1346}}} = \frac{-0.16}{0.0169} = -9.45$$

P-value = area under the z curve to the left of $-9.45 \approx 0$.

Since the P-value is less than α, H_0 is rejected. The sample data does support the conclusion that the proportion of undergraduate males who perceive pressure for a date as sexual harassment is lower than that of undergraduate females.

b Let π_1 denote the true proportion of graduate males who perceive pressure for a date as sexual harassment and π_2 denote the true proportion of graduate females who perceive pressure for a date as sexual harassment.

$H_0: \pi_1 - \pi_2 = 0$ $H_a: \pi_1 - \pi_2 < 0$

$\alpha = 0.05$

$$z = \frac{p_1 - p_2}{\sqrt{\dfrac{p_c(1-p_c)}{n_1} + \dfrac{p_c(1-p_c)}{n_2}}}$$

$n_1 = 565, \quad x_1 = 469, \quad p_1 = \dfrac{469}{565} = .8301, \quad n_2 = 575, \quad x_2 = 529, \quad p_2 = \dfrac{529}{575} = .9200$

$$p_c = \frac{n_1 p_1 + n_2 p_2}{n_1 + n_2} = \frac{469 + 529}{565 + 575} = 0.8754$$

$$z = \frac{(0.8301 - 0.9200)}{\sqrt{\dfrac{0.8754(0.1246)}{565} + \dfrac{0.8754(0.1246)}{575}}} = \frac{-0.0899}{0.0196} = -4.60$$

P-value = area under the z curve to the left of $-4.60 \approx 0$.

Since the P-value is less than α, H₀ is rejected. The sample data does support the conclusion that the proportion of graduate males who perceive pressure for a date as sexual harassment is lower than that of graduate females.

c   Let π_1 denote the true proportion of male graduate students who perceive pressure for a date as sexual harassment and let π_2 denote the true proportion male undergraduate students who perceive pressure for a date as sexual harassment.

$n_1 = 565, \quad x_1 = 469, \quad p_1 = \dfrac{469}{565} = .8301, \quad n_2 = 1336, \quad x_2 = 882, \quad p_2 = \dfrac{882}{1336} = .6602$

The 90% confidence interval for $\pi_1 - \pi_2$ is

$$(0.8301 - 0.6602) \pm 1.645 \sqrt{\frac{0.8301(0.1699)}{565} + \frac{0.6602(0.3398)}{1336}}$$

$\Rightarrow \quad 0.1699 \pm 1.6450 \quad (0.0204) \quad \Rightarrow \quad 0.1669 \pm 0.0336 \quad \Rightarrow \quad (0.1333, 0.2005)$.

Based on this interval, we believe that the proportion of male graduate students who perceive pressure for a date as sexual harassment is larger than that for male undergraduates. The difference in proportions may be as small as 0.1333 or as large as 0.2005.

11.77   Let μ_1 denote the mean self-esteem score of women at the University of Cincinnati and μ_2 denote the mean self-esteem score of women at the University of New Mexico.

H_0: $\mu_1 - \mu_2 = 0$ H_a: $\mu_1 - \mu_2 \neq 0$

$\alpha = 0.05$

Test statistic: $t = \dfrac{(\bar{x}_1 - \bar{x}_2) - 0}{\sqrt{\dfrac{s_1^2}{n_1} + \dfrac{s_2^2}{n_2}}}$

Assumptions: The two populations of self-esteem scores are normally distributed and the two samples are independently selected random samples.

$n_1 = 47$, $\bar{x}_1 = 32.55$, $s_1 = 4.41$, $n_2 = 24$, $\bar{x}_2 = 31.25$, $s_2 = 4.92$

$$t = \frac{(32.55 - 31.25)}{\sqrt{\dfrac{(4.41)^2}{47} + \dfrac{(4.92)^2}{24}}} = \frac{1.3}{1.19264} = 1.09$$

$$df = \frac{\left(\dfrac{s_1^2}{n_1} + \dfrac{s_2^2}{n_2}\right)^2}{\dfrac{1}{n_1 - 1}\left(\dfrac{s_1^2}{n_1}\right)^2 + \dfrac{1}{n_2 - 1}\left(\dfrac{s_2^2}{n_2}\right)^2} = \frac{(0.41379 + 1.0086)^2}{\dfrac{(0.41379)^2}{46} + \dfrac{(1.0086)^2}{23}} = 42.19$$

So $df = 42$ (rounded down to an integer)

P-value = 2(area under the 42 df t curve to the right of 1.09) $\approx$ 2(0.141) = 0.282.

Since the P-value exceeds α, the null hypothesis is not rejected. The data does not support the conclusion that the mean self-esteem scores of women differ at the two universities.

11.79 Let π_1 denote the true proportion of adults born deaf who remove the implants. Let π_2 denote the true proportion of adults who went deaf after learning to speak who remove the implants.

H_0: $\pi_1 - \pi_2 = 0$ H_a: $\pi_1 - \pi_2 \neq 0$

$\alpha = 0.01$

$$z = \frac{p_1 - p_2}{\sqrt{\dfrac{P_c(1-P_c)}{n_1} + \dfrac{P_c(1-P_c)}{n_2}}}$$

$n_1 = 250,\ x_1 = 75,\ p_1 = 0.3,\ n_2 = 250,\ x_2 = 25,\ p_2 = 0.1$

$$P_c = \frac{n_1 p_1 + n_2 p_2}{n_1 + n_2} = \frac{75 + 25}{250 + 250} = 0.2$$

$$z = \frac{(0.3 - 0.1)}{\sqrt{\dfrac{0.2(0.8)}{250} + \dfrac{0.2(0.8)}{250}}} = \frac{0.2}{0.03577} = 5.59$$

P-value = 2(area under the z curve to the right of 5.59) $\approx$ 0.

Since the P-value is less than α, the null hypothesis is rejected. The data does support the fact that the true proportion who remove the implants differs in those that were born deaf from that of those who went deaf after learning to speak.

Chapter 12

Exercises 12.1 – 12.13

12.1 **a** $0.020 < \text{P-value} < 0.025$

     b $0.040 < \text{P-value} < 0.045$

     c $0.035 < \text{P-value} < 0.040$

     d $\text{P-value} < 0.001$

     e $\text{P-value} > 0.100$

12.3 **a** df = 3 and $X^2 = 19.0$. From Appendix Table IX, P-value < 0.001. Since the P-value is less than α, H$_\circ$ is rejected.

     b If n = 40, then it is not advisable to use the chi-square test since one of the expected cell frequencies (cell corresponding to nut type 4) would be less than 5.

12.5 Let π_i denote the true proportion of bicycle accidents resulting in death. (i = Mon, Tues, Wed etc)

H$_\circ$: $\pi_{Su} = \pi_M = \pi_T = \pi_W = \pi_{Th} = \pi_F = \pi_{Sa} = \dfrac{1}{7}$

H$_a$:   H$_\circ$ is not true.

$\alpha = 0.05$

Test statistic: $X^2 = \sum \dfrac{(\text{observed count} - \text{expected count})^2}{\text{expected count}}$

Computations:

Day	Su	M	Tu	W	Th	F	Sa	Total
Frequency	14	13	12	15	14	17	15	100
Expected	14.286	14.286	14.286	14.286	14.286	14.286	14.286	

$$X^2 = \frac{(14 - 14.286)^2}{14.286} + \frac{(13 - 14.286)^2}{14.286} + \frac{(12 - 14.286)^2}{14.286} + \frac{(15 - 14.286)^2}{14.286}$$

$$+ \frac{(14-14.286)^2}{14.286} + \frac{(17-14.286)^2}{14.286} + \frac{(15-14.286)^2}{14.286} = 1.08$$

df = 6. From Appendix Table IX, P-value > 0.100. Since the P-value exceeds α, the null hypothesis is not rejected. From this data, it is not reasonable to conclude that the proportion of accidents is different for any of the days of the week.

12.7 Let π_i denote the true proportion of lottery ticket purchasers in age group i (i = 1, 2, 3).

H₀: $\pi_1 = .35,\ \pi_2 = .51,\ \pi_3 = .14$

Hₐ: at least one of the π_i is different from the hypothesized

$\alpha = 0.05$

Test statistic: $X^2 = \sum \dfrac{(\text{observed count} - \text{expected count})^2}{\text{expected count}}$

n = 36 + 130 + 34 = 200.

Expected count for each cell: 18–34: 200(.35) = 70, 35-64: 200(.51) = 102, 65+: 200(.14) = 28

None of the expected counts are less than 5. The article did not specify the method of sampling but we will assume that the 200 purchasers of the lottery tickets can be regarded as a random sample from the population.

$$X^2 = \frac{(36-70)^2}{70} + \frac{(130-102)^2}{102} + \frac{(34-28)^2}{28}$$

$$= 16.5143 + 7.6863 + 1.2857 = 25.4863$$

df = 2. From Appendix Table IX, P-value < 0.001. Since the P-value is less than α, H₀ is rejected. The data provide strong evidence to conclude that one or more of the three age groups buys a disproportionate share of lottery tickets.

White	679	0.507	507
Black	51	0.066	66
Hispanic	77	0.306	306
Asian	190	0.108	108
Other	3	0.013	13

$$X^2 = \frac{(679-507)^2}{507} + \frac{(51-66)^2}{66} + \frac{(77-306)^2}{306} + \frac{(190-108)^2}{108} + \frac{(3-13)^2}{13}$$

$$= 58.351 + 3.409 + 171.376 + 62.259 + 7.692 = 303.09$$

df =4. From Appendix Table IX, P-value < 0.001. At a significance level of $\alpha = 0.01$, H_o is rejected.

The data provide very strong evidence to conclude that the proportions of students graduating from colleges and universities in California differ from the respective proportions in the population.

12.9 Let π_1, π_2, π_3, π_4 denote the true proportions of homicides occurring during Winter, Spring, Summer, and Fall, respectively.

H_o: $\pi_1 = \pi_2 = \pi_3 = \pi_4 = 0.25$

H_a: H_o is not true.

$\alpha = 0.05$

Test statistic: $X^2 = \Sigma \dfrac{(\text{observed count} - \text{expected count})^2}{\text{expected count}}$

Computations: n = 1361

Season	Winter	Spring	Summer	Fall
Frequency	328	334	372	327
Expected	340.25	340.25	340.25	340.25

$$X^2 = \frac{(328-340.25)^2}{340.25} + \frac{(334-340.25)^2}{340.25} + \frac{(372-340.25)^2}{340.25} + \frac{(327-340.25)^2}{340.25}$$
$$= 0.4410 + 0.1148 + 2.9627 + 0.5160 = 4.03453.$$

df = 3. From Appendix Table IX, P-value > 0.100. Since the P-values exceeds α, the null hypothesis is not rejected. The data collected does not suggest that there is a difference in the proportion of homicides occurring in the four seasons.1

12.11 **a**   Let π_i denote the proportion for phenotype i (i = 1, 2, 3).

H_o: $\pi_1 = 0.25$, $\pi_2 = 0.5$, $\pi_3 = 0.25$

H_a: H_o is not true.

$\alpha = 0.05$

Test statistic: $X^2 = \sum \dfrac{(\text{observed count} - \text{expected count})^2}{\text{expected count}}$

Computations: df = 2 and the computed X² value is 4.63. From Appendix Table IX, P-value > 0.1. Since the P-value is not less than α, the null hypothesis is not rejected. The data do not contradict the researcher's theory.

b   The analysis and conclusion would remain the same. The sample size is used only to calculate the expected cell frequencies. It has no influence on the degrees of freedom, or the P-value. (It does improve the fit of the χ^2 distribution to the sampling distribution of the test statistic.)

12.13   Let π_i denote the proportion of homing pigeons who prefer direction i (i = 1, 2, 3, 4, 5, 6, 7, 8).

H₀: $\pi_1 = \pi_2 = \pi_3 = \pi_4 = \pi_5 = \pi_6 = \pi_7 = \pi_8 = \dfrac{1}{8}$

Hₐ: H₀ is not true.

α = 0.10

Test statistic: $X^2 = \sum \dfrac{(\text{observed count} - \text{expected count})^2}{\text{expected count}}$

Computations:

Direction	1	2	3	4	5	6	7	8	Total
Frequency	12	16	17	15	13	20	17	10	120
Expected	15	15	15	15	15	15	15	15	

$$X^2 = \frac{(12-15)^2}{15} + \frac{(16-15)^2}{15} + \frac{(17-15)^2}{15} + \frac{(15-15)^2}{15}$$
$$+ \frac{(13-15)^2}{15} + \frac{(20-15)^2}{15} + \frac{(17-15)^2}{15} + \frac{(10-15)^2}{15} = \frac{72}{15} = 4.8$$

df = 7. From Appendix Table IX, P-value > 0.100. Since the P-value exceeds α, the null hypothesis is not rejected. The data supports the hypothesis that when homing pigeons are disoriented in a certain manner, they exhibit no preference for any direction of flight after take-off.

Exercises 12.14 – 12.34

12.15   **a** $\alpha = 0.10$, df $= (4-1)(5-1) = 12$, and $X^2 = 7.2$. From Appendix Table IX, P-value > 0.10. Since the P-value exceeds α, the null hypothesis would not be rejected. The data are consistent with the hypothesis that educational level and preferred candidate are independent factors.

     b $\alpha = 0.05$, df $= (4-1)(4-1) = 9$, and $X^2 = 14.5$. From Appendix Table IX, P-value > 0.100. Since the P-value exceeds α, the null hypothesis would not be rejected. The data are consistent with the hypothesis that educational level and preferred candidate are independent factors.

12.17 H_o: City of residence and type of vehicle used most often are independent
       H_a: City of residence and type of vehicle used most often are not independent

       $\alpha = 0.05$

       Test statistic: $X^2 = \sum \dfrac{(\text{observed count} - \text{expected count})^2}{\text{expected count}}$

       Observed and expected frequencies are given in the table below (expected frequencies in parentheses)

	Concord	Pleasant Hills	North San Fran.	
Small	68 (89.06)	83 (107.02)	221 (175.92	372
Compact	63 (56.74)	68 (68.18)	106 (112.08)	237
Midsize	88 (84.51)	123 (101.55)	142 (166.94)	353
Large	24 (12.69)	18 (15.25)	11 (25.06)	53
	243	292	480	1015

$$X^2 = \frac{(68-89.06)^2}{89.06} + \frac{(83-107.02)^2}{107.02} + \frac{(221-175.92)^2}{175.92} + \frac{(63-56.74)^2}{56.74}$$

$$+ \frac{(68-68.18)^2}{68.18} + \frac{(106-112.08)^2}{112.08} + \frac{(88-84.51)^2}{84.51} + \frac{(123-101.55)^2}{101.55} + \frac{(142-166.94)^2}{166.94} + \frac{(24-12.69)^2}{12.69}$$

$$+ \frac{(18-15.25)^2}{15.25} + \frac{(11-25.06)^2}{25.06} = 49.813$$

df = (4 – 1)(3 – 1) = 6. From Appendix Table IX, P-value < 0.001. Hence the null hypothesis is rejected. There is enough evidence to suggest that there is an association between the city of residence and the type of vehicle that is used most often.

12.19 a   The table below gives the row percentages for each smoking category.

	< 1/wk	1/wk	2-4/wk	5-6/wk	1/day
Never smoked	33.00	15.79	22.42	11.17	17.62
Smoked in the past	18.53	12.41	23.22	14.68	31.17
Currently smokes	21.58	11.90	19.36	12.19	34.97

The proportions falling into each category appear to be dissimilar. For instance, only 17.62% of the subjects in the "never smoked" category consumed one drink per day, whereas 34.97% of those in the "currently smokes" category consume one drink per day. Similar discrepancies are seen for other categories as well.

b H_0: Smoking status and alcohol consumption are independent.
H_a: Smoking status and alcohol consumption are not independent.

$\alpha = 0.05$ (No significance level is not given in the problem. We use 0.05 for illustration.)

Test statistic: $X^2 = \sum \frac{(\text{observed count} - \text{expected count})^2}{\text{expected count}}$

Observed and expected frequencies are given in the table below (expected frequencies in parentheses)

<table>
<tr><td></td><td></td><td colspan="5" align="center">Alcohol consumption
(no. of drinks)</td><td></td></tr>
<tr><td></td><td></td><td><1/wk</td><td>1/wk</td><td>2-4/wk</td><td>5-6/wk</td><td>1/day</td><td></td></tr>
<tr><td></td><td>Never
Smoked</td><td>3577
(2822.22)</td><td>1711
(1520.12)</td><td>2430
(2427.33)</td><td>1211
(1372.96)</td><td>1910
(2696.37)</td><td>10839</td></tr>
<tr><td>Smoking
status</td><td>Smoked in
the past</td><td>1595
(2241.58)</td><td>1068
(1207.37)</td><td>1999
(1927.93)</td><td>1264
(1090.49)</td><td>2683
(2141.62)</td><td>8609</td></tr>
<tr><td></td><td>Currently
Smokes</td><td>524
(632.19)</td><td>289
(340.51)</td><td>470
(543.74)</td><td>296
(307.55)</td><td>849
(604.00)</td><td>2428</td></tr>
<tr><td></td><td></td><td>5696</td><td>3068</td><td>4899</td><td>2771</td><td>5442</td><td>21876</td></tr>
</table>

$$X^2 = \frac{(3577-2822.22)^2}{2822.22} + \frac{(1711-1520.12)^2}{1520.12} + \frac{(2430-2427.33)^2}{2427.33} + \frac{(1211-1372.96)^2}{1372.96}$$

$$+\frac{(1910-2696.37)^2}{2696.37} + \frac{(1595-2241.58)^2}{2241.58} + \frac{(1068-1207.37)^2}{1207.37} + \frac{(1999-1927.93)^2}{1927.93}$$

$$+\frac{(1264-1090.49)^2}{1090.49} + \frac{(2683-2141.62)^2}{2141.62} + \frac{(524-632.19)^2}{632.19} + \frac{(289-340.51)^2}{340.51}$$

$$+\frac{(470-543.74)^2}{543.74} + \frac{(296-307.55)^2}{307.55} + \frac{(849-604.00)^2}{604.00} = 980.068$$

df = (3 − 1)(5 − 1) = 8. From Appendix Table IX, P-value < 0.001. Hence the null hypothesis is rejected. The data provide strong evidence to conclude that smoking status and alcohol consumption are not independent.

c   The result of the test in part **b** is consistent with our observations in part **a**.

12.21 H₀: There is no dependence (i.e., independent) between handgun purchase within the year prior to death and whether or not the death was a suicide.

   Hₐ: There is a dependence between handgun purchase within the year prior to death and whether or not the death was a suicide.

$\alpha = 0.05$ (No significance level is given in the problem. We use 0.05 for illustration.)

Test statistic: $X^2 = \sum \dfrac{(\text{observed count } - \text{ expected count})^2}{\text{expected count}}$

Observed and expected frequencies are given in the table below (expected frequencies in parentheses)

	Suicide	Not suicide	
Purchased Handgun	4 (0.27)	12 (15.73)	16
No handgun purchase	63 (66.73)	3921 (3917.27)	3984
	67	3933	4000

$$X^2 = \frac{(4-0.27)^2}{0.27} + \frac{(63-66.73)^2}{66.73} + \frac{(12-15.73)^2}{15.73} + \frac{(3921-3917.27)^2}{3917.27} = 53.067$$

df = (2 – 1)(2 – 1) = 1. From Appendix Table IX, P-value < 0.001. Hence the null hypothesis is rejected.

The data provide strong evidence to conclude that there is an association between handgun purchase within the year prior to death and whether or not the death was a suicide.

NOTE: One cell has an expected count that is less than 5. The chi-square approximation is probably not satisfactory.

12.23 H_o: Position and Role are independent.
H_a: Position and Role are not independent.

$\alpha = 0.01$

Test statistic: $X^2 = \sum \dfrac{(\text{observed count} - \text{expected count})^2}{\text{expected count}}$

Observed and expected frequencies are given in the table below (expected frequencies in parentheses)

	Initiate chase	Participate in chase	
Center position	28 (39.04)	48 (36.96)	76
Wing position	66 (54.96)	41 (52.04)	107
	94	89	183

$$X^2 = \frac{(28-39.04)^2}{39.04} + \frac{(66-54.96)^2}{54.96} + \frac{(48-36.96)^2}{36.96} + \frac{(41-52.04)^2}{52.04} = 10.97$$

df = (2 − 1)(2 − 1) = 1. From Appendix Table IX, P-value < 0.001. Hence the null hypothesis is rejected. The data provide strong evidence to conclude that there is an association between position and role.

For the chi-square analysis to be valid, the observations on the 183 lionesses in the sample are assumed to be independent.

12.25 H_o: There is no dependence between response and region of residence.
H_a: There is a dependence between response and region of residence.

$\alpha = 0.01$

Test statistic: $X^2 = \sum \dfrac{(\text{observed count} - \text{expected count})^2}{\text{expected count}}$

		Response		
		Agree	Disagree	
	Northeast	130	59	189
		(150.35)	(38.65)	
Region	West	146	42	188
		(149.55)	(38.45)	
	Midwest	211	52	263
		(209.22)	(53.78)	
	South	291	47	338
		(268.88)	(69.12)	
		778	200	978

$$X^2 = \frac{(130-150.35)^2}{150.35} + \frac{(59-38.65)^2}{38.65} + \frac{(146-149.55)^2}{149.55} + \frac{(42-38.45)^2}{38.45}$$

$$+ \frac{(211-209.22)^2}{209.22} + \frac{(52-53.78)^2}{53.78} + \frac{(291-268.88)^2}{268.88} + \frac{(47-69.12)^2}{69.12}$$

$= 2.754 + 10.714 + 0.084 + 0.329 + 0.015 + 0.059 + 1.820 + 7.079 = 22.855$

df = (4 − 1)(2 − 1) = 3. From Appendix Table IX, 0.001 > P-value. Since the P-value is less than α, the null hypothesis is rejected. The data supports the conclusion that there is a dependence between response and region of residence.

12.27 H_o: The proportion of correct sex identifications is the same for each nose view.
H_a: The proportion of correct sex identifications is not the same for each nose view.
$\alpha = 0.05$

Test statistic: $X^2 = \sum \dfrac{(\text{observed count } - \text{ expected count})^2}{\text{expected count}}$

		Nose view			
		Front	Profile	Three quarter	
Sex ID	Correct	23 (26)	26 (26)	29 (26)	78
	Not Correct	17 (14)	14 (14)	11 (14)	42
		40	40	40	120

$$X^2 = \frac{(23-26)^2}{26} + \frac{(26-26)^2}{26} + \frac{(29-26)^2}{26} + \frac{(17-14)^2}{14} + \frac{(14-14)^2}{14} + \frac{(11-14)^2}{14}$$

$= 0.346 + 0.000 + 0.346 + 0.643 + 0.000 + 0.643 = 1.978$

df = (2 – 1)(3 – 1) = 2. From Appendix Table IX, P-value > 0.10. Since the P-value exceeds α, the null hypothesis is not rejected. The data does not support the hypothesis that the proportions of correct sex identifications differ for the three different nose views.

12.29   H₀: the response category proportions are the same for the two cover designs.

   Hₐ: the proportions are not the same for all response categories for the two cover designs.

$\alpha = 0.05$

Test statistic: $X^2 = \sum \dfrac{(\text{observed count } - \text{ expected count})^2}{\text{expected count}}$

	1 - 7	8 - 14	15 – 31	32 - 60	Not returned	
Graphic	70 (75.8)	76 (63.5)	51 (50.2)	19 (25.1)	198 (199.4)	414
Plain	84 (78.2)	53 (65.5)	51 (51.8)	32 (25.9)	207 (205.6)	427
	154	129	102	51	405	841

Since all expected cell counts are at least 5, the X^2 statistics can be used.

$df = (2-1)(5-1) = 4$

$$X^2 = \frac{(70-75.8)^2}{75.8} + \frac{(76-63.5)^2}{63.5} + \frac{(51-50.2)^2}{50.2} + \frac{(19-25.1)^2}{25.1} + \frac{(198-199.4)^2}{199.4}$$

$$+ \frac{(84-78.2)^2}{78.2} + \frac{(53-65.5)^2}{65.5} + \frac{(51-51.8)^2}{51.8} + \frac{(32-25.9)^2}{25.9} + \frac{(207-205.6)^2}{205.6}$$

$$= 0.4438 + 2.4606 + 0.0127 + 1.4825 + 0.0098 + 0.4302 + 2.3855 + 0.0124 + 1.4367 + 0.0095 = 8.6837$$

From Appendix Table IX, $0.065 < \text{P-value} < 0.070$. Since the P-value exceeds α, the null hypothesis is not rejected. The data does not support the theory that the proportions falling in the various response categories differ for the two cover designs. There is no evidence that cover design affects speed of response.

12.31 H$_0$: There is no dependence between the approach used and whether or not a donation is obtained.

H$_a$: There is a dependence between the approach used and whether or not a donation is obtained.

$\alpha = 0.05$ (No significance level is given in the problem. We use 0.05 for illustration.)

Test statistic: $X^2 = \sum \dfrac{(\text{observed count} - \text{expected count})^2}{\text{expected count}}$

Observed and expected frequencies are given in the table below (expected frequencies in parentheses)

	Contribution made	No contribution made	
Picture of a smiling child	18 (18.33)	12 (11.67)	30
Picture of an unsmiling child	14 (18.33)	16 (11.67)	30
Verbal message	16 (15.89)	10 (10.11)	26
Identification of charity only	18 (13.44)	4 (8.56)	22
	66	42	108

$$X^2 = \frac{(18-18.33)^2}{18.33} + \frac{(14-18.33)^2}{18.33} + \frac{(16-15.89)^2}{15.89} + \frac{(18-13.44)^2}{13.44} + \frac{(12-11.67)^2}{11.67}$$

$$+ \frac{(16-11.67)^2}{11.67} + \frac{(10-10.11)^2}{10.11} + \frac{(4-8.56)^2}{8.56} = 6.621$$

df = (4 – 1)(2 – 1) = 3. From Appendix Table IX, P-value > 0.05. Hence the null hypothesis is not rejected. The data do not provide sufficient evidence to conclude that there is a dependence between the approach used to obtain donations and whether or not a donation is successfully obtained.

12.33 H₀: Job satisfaction and teaching level are independent.
H_a: Job satisfaction and teaching level are dependent.

$\alpha = 0.05$

Test statistic: $X^2 = \sum \dfrac{(\text{observed count } - \text{ expected count})^2}{\text{expected count}}$

Computations:

		Satisfied	Unsatisfied	
	College	74 (63.763)	43 (53.237)	117
Teaching Level	High School	224 (215.270)	171 (179.730)	395
	Elementary	126 (144.967)	140 (121.033)	266
	Total	424	354	778

Job satisfaction

$$X^2 = \frac{(74-63.763)^2}{63.763} + \frac{(43-53.237)^2}{53.237} + \frac{(224-215.270)^2}{215.270}$$

$$+ \frac{(171-179.730)^2}{179.730} + \frac{(126-144.967)^2}{144.967} + \frac{(140-121.023)^2}{121.023}$$

$$= 1.644 + 1.968 + 0.354 + 0.424 + 2.482 + 2.972 = 9.844$$

df = (3 – 1)(2 – 1) = 2. From Appendix Table IX, 0.010 > P-value > 0.005. Since the P-value is less than α, H₀ is rejected. The data supports the conclusion that there is a dependence between job satisfaction and teaching level.

Exercises 12.35 – 12.44

12.35   **a**   Let π_i denote the true proportion of policy holders in Astrological sign group i. (i = 1 for Aquarius, i = 2 for Aries, ..., i = 12 for Virgo).

H_0: $\pi_1 = \pi_2 = \pi_3 = \pi_4 = \pi_5 = \pi_6 = \pi_7 = \pi_8 = \pi_9 = \pi_{10} = \pi_{11} = \pi_{12} = 1/12$

H_a: at least one of the true proportions differ from 1/12

$\alpha = 0.05$ (A significance level is not specified in this problem. We use $\alpha = 0.05$ for illustration.)

Test statistic: $X^2 = \sum \dfrac{(\text{observed count } - \text{ expected count})^2}{\text{expected count}}$

n = 35,666 + 37,926 + 38,126 + 54,906 + 37,179 + 37,354 + 37,910 + 36,677 + 34,175 + 35,352 + 37,179 + 37,718 = 460,168

Expected count for each cell = 460,168(1/12) = 38,347.3.

$$X^2 = \frac{(35,666 - 38,347.3)^2}{38,347.3} + \frac{(37,926 - 38,347.3)^2}{38,347.3} + \frac{(38,126 - 38,347.3)^2}{38,347.3} + \frac{(54,906 - 38,347.3)^2}{38,347.3}$$

$$+ \frac{(37,179 - 38,347.3)^2}{38,347.3} + \frac{(37,354 - 38,347.3)^2}{38,347.3} + \frac{(37,910 - 38,347.3)^2}{38,347.3} + \frac{(36,677 - 38,347.3)^2}{38,347.3}$$

$$+ \frac{(34,175 - 38,347.3)^2}{38,347.3} + \frac{(35,352 - 38,347.3)^2}{38,347.3} + \frac{(37,179 - 38,347.3)^2}{38,347.3} + \frac{(37,718 - 38,347.3)^2}{38,347.3}$$

=187.48 + 4.63 + 1.28 + 7150.16 + 35.60 + 25.73 + 4.99 + 72.76 + 453.97 + 233.97 + 35.60 + 10.33 = 8,216.48

df =11. From Appendix Table IX, P-value < 0.001. Since the P-value is less than than α, H_0 is rejected. The data provide very strong evidence to conclude that the proportions of policy holders are not all equal for the twelve astrological signs.

b   Sign of Capricorn covers birthdates between December 22 and January 20 which is the summer season in Australia. One possible explanation for the higher than expected proportion of policy holders for this sign might be that more teenagers start driving during the summer months than any other months and hence more policies are issued during this period.

c   Let π_i denote the true proportion of policy holders in Astrological sign group i who make claims (i = 1 for Aquarius, i = 2 for Aries, ..., i = 12 for Virgo).

H_o: $\pi_1 = 35666/460168 = 0.077506$, $\pi_2 = 37926/460168 = 0.082418$,
$\pi_3 = 38126/460168 = 0.082852$, $\pi_4 = 54906/460168 = 0.119317$,
$\pi_5 = 37179/460168 = 0.080794$, $\pi_6 = 37354/460168 = 0.081175$,
$\pi_7 = 37910/460168 = 0.082383$, $\pi_8 = 36677/460168 = 0.079703$,
$\pi_9 = 34175/460168 = 0.074266$, $\pi_{10} = 35352/460168 = 0.076824$,
$\pi_{11} = 37179/460168 = 0.080794$, $\pi_{12} = 37718/460168 = 0.081966$

H_a: at least one of the true proportions differs from the hypothesized value.

$\alpha = 0.01$ (A significance level is not specified in this problem. We use $\alpha = 0.01$ for illustration.)

Test statistic: $X^2 = \sum \dfrac{(\text{observed count} - \text{expected count})^2}{\text{expected count}}$

$n = 1000$

The required calculations for obtaining the expected cell frequencies are summarized in the table below. The number of policy holders of the company for the different astrological signs is given in the second column of the table. The corresponding proportions are given in the third column. The number of policy holders in the sample making claims for the different astrological signs is given in column 4. The corresponding expected number of claims is given in the last column. The expected number is calculated as follows:

$$\left(\begin{array}{c} \text{Expected number of claims} \\ \text{for this astrological sign} \end{array} \right) = 1000 \times \frac{\text{Number of policy holders with this sign}}{\text{Total number of policy holders}}$$

246

Sign	Number of policy holders	Proportion of policy holders	Number of claims in the sample	Expected number of claims in the sample
Aquarius	35666	0.077506	85	77.506
Aries	37926	0.082418	83	82.418
Cancer	38126	0.082852	82	82.852
Capricorn	54906	0.119317	88	119.317
Gemini	37179	0.080794	83	80.794
Leo	37354	0.081175	83	81.175
Libra	37910	0.082383	83	82.383
Pisces	36677	0.079703	82	79.703
Sagittarius	34175	0.074266	81	74.266
Scorpio	35352	0.076824	85	76.824
Taurus	37179	0.080794	84	80.794
Virgo	37718	0.081966	81	81.966

$$X^2 = \frac{(85-77.506)^2}{77.506} + \frac{(83-82.418)^2}{82.418} + \frac{(82-82.852)^2}{82.852} + \frac{(88-119.317)^2}{119.317}$$

$$+ \frac{(83-80.794)^2}{80.794} + \frac{(83-81.175)^2}{81.175} + \frac{(83-82.383)^2}{82.383} + \frac{(82-79.703)^2}{79.703}$$

$$+ \frac{(81-74.266)^2}{74.266} + \frac{(85-76.824)^2}{76.824} + \frac{(84-80.794)^2}{80.794} + \frac{(81-81.966)^2}{81.966}$$

$= 0.72449 + 0.00411 + 0.00877 + 8.21987 + 0.06021 + 0.04104 + 0.00462 + 0.06617 + 0.61053$

$\quad + 0.87011 + 0.12718 + 0.01138 = 10.7485$

df =11. From Appendix Table IX, P-value > 0.45. At a significance level of $\alpha = 0.01$, H_0 cannot be rejected. It cannot be rejected even at a significance level of $\alpha = 0.05$. The data do not provide evidence to conclude that the proportions of claims are consistent with the proportions of policy holders for the various astrological signs. However, it is worth noting that the proportion of claims by policy holders in the sample belonging to Capricorn is much smaller than what would be expected based on the overall proportion of policy holders with this sign!

12.37 H_0: Age at death and location of death are independent.
H_a: Age at death and location of death are dependent.

$\alpha = 0.01$

Test statistic: $X^2 = \sum \dfrac{(\text{observed count} - \text{expected count})^2}{\text{expected count}}$

Computations:

Age	Location Home	Acute care	Chronic care	Total
15 - 54	94	418	23	535
	(90.2)	(372.5)	(72.3)	
55 - 64	116	524	34	674
	(113.6)	(469.3)	(91.1)	
65 - 74	156	581	109	846
	(142.7)	(589)	(114.3)	
Over 74	138	558	238	934
	(157.5)	(650.3)	(126.2)	
Total	504	2081	404	2989

$$X^2 = \frac{(94-90.2)^2}{90.2} + \frac{(418-372.5)^2}{372.5} + \frac{(23-72.3)^2}{72.3} + \frac{(116-113.6)^2}{113.6}$$

$$+ \frac{(524-469.3)^2}{469.3} + \frac{(34-91.1)^2}{91.1} + \frac{(156-142.7)^2}{142.7} + \frac{(581-589)^2}{589}$$

$$+ \frac{(109-114.3)^2}{114.3} + \frac{(138-157.5)^2}{157.5} + \frac{(558-650.3)^2}{650.3} + \frac{(238-126.2)^2}{126.2}$$

$$= 0.16 + 5.56 + 33.63 + 0.05 + 6.39 + 35.79 + 1.25 + 0.11 + 0.25 + 2.41 + 13.09 + 98.94$$

$$= 197.62$$

df = (4 – 1)(3 – 1) = 6. From Appendix Table IX, P-value < 0.001. Since the P-value is less than α, the null hypothesis is rejected. The data strongly suggests that the variables, age at death and location of death, are dependent.

12.39 H_o: Sex and relative importance assigned to work and home are independent.
H_a: Sex and relative importance assigned to work and home are dependent.

$\alpha = 0.05$

Test statistic: $X^2 = \sum \dfrac{(\text{observed count} - \text{expected count})^2}{\text{expected count}}$

Computations:

Relative Importance

	work > home	work = home	work < home	Total
Female	68 (79.3)	26 (25.0)	94 (83.7)	188
Male	75 (63.7)	19 (20.0)	57 (67.3)	151
Total	143	45	151	339

$X^2 = 1.61 + 0.04 + 1.26 + 2.01 + 0.05 + 1.56 = 6.54$

df $= (2 - 1)(3 - 1) = 2$. From Appendix Table IX, $0.040 > \text{P-value} > 0.035$. Since the P-value is less than α, the null hypothesis is rejected. The data suggests that sex and relative importance assigned to work and home are dependent.

12.41 H_0:   Age and "Rate believed attainable" are independent.

      H_a: Age and "Rate believed attainable" are not dependent.

$\alpha = 0.01$

Test statistic: $X^2 = \sum \dfrac{(\text{observed count} - \text{expected count})^2}{\text{expected count}}$

Computations:

Rates believed attainable

		0 - 5	6 - 10	11 - 15	Over 15	Total
	Under 45	15 (28.4)	51 (72.1)	51 (28.2)	29 (17.3)	146
	45 - 54	31 (54.8)	133 (139.3)	70 (54.5)	48 (33.4)	282
Age	55 - 64	59 (49.2)	139 (124.9)	35 (48.9)	20 (29.9)	253
	65 over	84 (56.6)	157 (143.7)	32 (56.3)	18 (34.4)	291
	Total	189	480	188	115	972

$$X^2 = \frac{(15-28.4)^2}{28.4} + \frac{(51-72.1)^2}{72.1} + \frac{(51-28.2)^2}{28.2} + \frac{(29-17.3)^2}{17.3}$$

$$+ \frac{(31-54.8)^2}{54.8} + \frac{(133-139.3)^2}{139.3} + \frac{(70-54.5)^2}{54.5} + \frac{(48-33.4)^2}{33.4}$$

$$+ \frac{(59-49.2)^2}{49.2} + \frac{(139-124.9)^2}{124.9} + \frac{(35-48.9)^2}{48.9} + \frac{(20-29.9)^2}{29.9}$$

$$+ \frac{(84-56.6)^2}{56.6} + \frac{(157-143.7)^2}{143.7} + \frac{(32-56.3)^2}{56.3} + \frac{(18-34.4)^2}{34.4}$$

$$= 6.31 + 6.17 + 18.35 + 7.96 + 10.36 + 0.28 + 4.38 + 6.42$$

$$+ 1.95 + 1.58 + 3.97 + 3.3 + 13.28 + 1.23 + 10.48 + 7.84$$

$$= 103.87$$

df = $(4-1)(4-1) = 9$. From Appendix Table IX, $0.001 >$ P-value. Since the P-value is less than α, the null hypothesis is rejected. The data very strongly suggests that the variables, age and "Rate believed attainable" are dependent.

12.43 H_o: The true proportions of individuals in each of the cocaine use categories do not differ for the three treatments.

   H_a: The true proportions of individuals in each of the cocaine use categories differ for the three treatments.

$\alpha = 0.05$

Test statistic: $X^2 = \sum \dfrac{(\text{observed count} - \text{expected count})^2}{\text{expected count}}$

Computations:

Treatment

Usage	A	B	C	Total
None	149 (118.9)	75 (84.8)	8 (28.3)	232
1 - 2	26 (34.8)	27 (24.9)	15 (8.3)	68
3 - 6	6 (19.0)	20 (13.5)	11 (4.5)	37
7 or more	4 (12.3)	10 (8.8)	10 (2.9)	24
Total	185	132	44	361

$$X^2 = \frac{(149-118.9)^2}{118.9} + \frac{(75-84.8)^2}{84.8} + \frac{(8-28.3)^2}{28.3} + \frac{(26-34.8)^2}{34.8}$$

$$+ \frac{(27-24.9)^2}{24.9} + \frac{(15-8.3)^2}{8.3} + \frac{(6-19)^2}{19} + \frac{(20-13.5)^2}{13.5}$$

$$+ \frac{(11-4.5)^2}{4.5} + \frac{(4-12.3)^2}{12.3} + \frac{(10-8.8)^2}{8.8} + \frac{(10-2.9)^2}{2.9}$$

$= 7.62 + 1.14 + 14.54 + 2.25 + 0.18 + 5.44 + 8.86 + 3.1 + 9.34 + 5.6 + 0.17 + 17.11 = 75.35$

df = (4 – 1)(3 – 1) = 6. From Appendix Table IX, 0.001 > P-value. Since the P-value is less than α, the null hypothesis is rejected.

Note: There are two cells with expected counts of 5 or less. If you combine the usage category "3 – 6" with "7 or more," the following analysis results.

Treatment

Usage	A	B	C	Total
none	149 (118.9)	75 (84.8)	8 (28.3)	232
1 - 2	26 (34.8)	27 (24.9)	15 (8.3)	68
3 or more	10 (31.3)	30 (22.3)	21 (7.4)	61
Total	185	132	44	361

$$X^2 = \frac{(149-118.9)^2}{118.9} + \frac{(75-84.8)^2}{84.8} + \frac{(8-28.3)^2}{28.3}$$

$$+ \frac{(26-34.8)^2}{34.8} + \frac{(27-24.9)^2}{24.9} + \frac{(15-8.3)^2}{8.3}$$

$$+ \frac{(10-31.3)^2}{31.3} + \frac{(30-22.3)^2}{22.3} + \frac{(21-7.4)^2}{7.4}$$

$= 7.62 + 1.14 + 14.54 + 2.25 + 0.18 + 5.44 + 14.46 + 2.65 + 24.75 = 73.03$

df = (3 - 1)(3 - 1) = 4. From Appendix Table IX, 0.001 > P-value. Since the P-value is less than α, the null hypothesis is rejected.

In either analysis, the null hypothesis is rejected. The data suggests the true proportions of individuals in each of the cocaine use categories differ for the three treatments.

Chapter 13

Exercises 13.1 – 13.11

13.1 **a** $y = -5.0 + 0.017x$

      b   When $x = 1000$, $y = -5 + 0.017(1000) = 12$
           When $x = 2000$, $y = -5 + 0.017(2000) = 29$

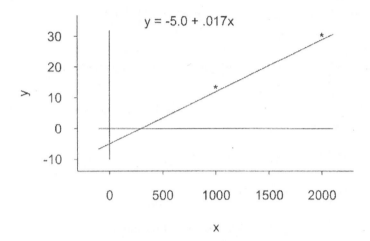

      c   When $x = 2100$, $y = -5 + (0.017)(2100) = 30.7$

      d   0.017

      e $0.017(100) = 1.7$

      f   It is stated that the community where the given regression model is valid has no
           small houses. Therefore, there is no assurance that the model is adequate for
           predicting usage based on house size for small houses. Consequently, it is not
           advisable to use the model to predict usage for a 500 sqft. house.

13.3 **a**   The mean value of serum manganese when Mn intake is 4.0 is $-2 + 1.4(4) = 3.6$.
           The mean value of serum manganese when Mn intake is 4.5 is $-2 + 1.4(4.5) = 4.3$.

      b $\dfrac{5 - 3.6}{1.2} = 1.17$

           $P(\text{serum Mn over } 5) = P(1.17 < z) = 1 - 0.8790 = 0.121$.

c   The mean value of serum manganese when Mn intake is 5 is $-2 + 1.4(5) = 5$.

$$\frac{5-5}{1.2} = 0, \qquad \frac{3.8-5}{1.2} = -1$$

P(serum Mn over 5) = $P(0 < z) = 0.5$
P(serum Mn below 3.8) = $P(z < -1) = 0.1587$

13.5   **a**   The expected change in price associated with one extra square foot of space is 47. The expected change in price associated with 100 extra square feet of space is $47(100) = 4700$.

        b   When $x = 1800$, the mean value of y is $23000 + 47(1800) = 107600$.

$$\frac{110000 - 107600}{5000} = 0.48 \qquad \frac{100000 - 107600}{5000} = -1.52$$

$P(y > 110000) = P(0.48 < z) = 1 - 0.6844 = 0.3156$

$P(y < 100000) = P(z < -1.52) = 0.0643$

Approximately 31.56% of homes with 1800 square feet would be priced over 110,000 dollars and about 6.43% would be priced under 100,000 dollars.

13.7   **a** $r^2 = 1 - \dfrac{\text{SSResid}}{\text{SSTo}} = 1 - \dfrac{27.890}{73.937} = 1 - 0.3772 = 0.6228$

        b $s_e^2 = \dfrac{\text{SSResid}}{n-2} = 1 - \dfrac{27.890}{13-2} = \dfrac{27.890}{11} = 2.5355$

          $s_e = \sqrt{2.5355} = 1.5923$

The magnitude of a typical deviation of residence half-time (y) from the population regression line is estimated to be about 1.59 hours.

        c $b = 3.4307$

        d $\hat{y} = 0.0119 + 3.4307(1) = 3.4426$

13.9   **a** $r^2 = 1 - \dfrac{2620.57}{22398.05} = 0.883$

b $s_e = \sqrt{\dfrac{2620.57}{14}} = \sqrt{187.184} = 13.682$ with 14 d f.

13.11   **a**

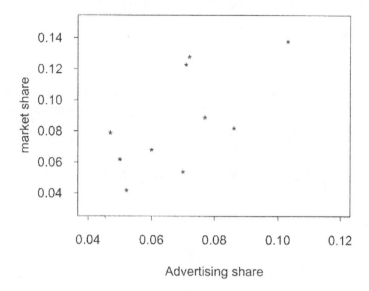

There seems to be a general tendency for y to increase at a constant rate as x increases. However, there is also quite a bit of variability in the y values. It is questionable whether a simple linear regression model using x is adequate for predicting. It may be advisable to look for additional predictor variables.

b   Summary values are: $n = 10$, $\Sigma x = 0.688$, $\Sigma x^2 = 0.050072$, $\Sigma y = 0.835$, $\Sigma y^2 = 0.079491$, $\Sigma xy = 0.060861$.

$$b = \frac{0.060861 - \left[\dfrac{(0.688)(0.835)}{10}\right]}{0.050072 - \left[\dfrac{(0.688)^2}{10}\right]} = \frac{0.003413}{0.0027376} = 1.246712$$

$a = 0.0835 - (1.246712)(0.0688) = -0.002274$

The equation of the estimated regression line is $\hat{y} = -0.002274 + 1.246712x$.

The predicted market share, when advertising share is 0.09, would be $-0.002274 + 1.246712(0.09) = 0.10993$.

c
$$SST_o = 0.079491 - \left[\frac{(0.835)^2}{10}\right] = 0.0097685$$

SSResid = 0.079491 − (−0.002274)(0.835) − (1.246712)(0.060861) = 0.0055135

$$r^2 = 1 - \frac{0.0055135}{0.0097685} = 1 - 0.564 = 0.436$$

This means that 43.6% of the total variability in market share (y) can be explained by the simple linear regression model relating market share and advertising share (x).

d
$$s_e = \sqrt{\frac{0.0055135}{8}} = \sqrt{0.000689} = 0.0263 \text{ with } 8 \, d\,f.$$

Exercises 13.12 – 13.26

13.13   **a** $\sum(x-\bar{x})^2 = 250$ $\sigma_b = \dfrac{4}{\sqrt{250}} = 0.253$

b $\sum(x-\bar{x})^2 = 500$ $\sigma_b = \dfrac{4}{\sqrt{500}} = 0.179$

No, the resulting value of σ_b is not half of what it was in **a**. However, the resulting value of σ_b^2 is half of what it was in **a**.

c   It would require 4 observations at each x value to yield a value of σ_b which is half the value calculated in **a**. In this case $\sum(x-\bar{x})^2 = 1000$, so
$$\sigma_b = \frac{4}{\sqrt{1000}} = 0.1265$$

13.15   **a** $s_e = \sqrt{\dfrac{1235.47}{13}} = \sqrt{95.036} = 9.7486$

$$s_b = \frac{9.7486}{\sqrt{4024.2}} = \frac{9.7486}{63.4366} = 0.1537$$

b   The 95% confidence interval for β is $2.5 \pm (2.16)(0.1537) \Rightarrow 2.5 \pm 0.33 \Rightarrow (2.17, 2.83)$.

c   The interval is relatively narrow. However, whether β has been precisely estimated or not depends on the particular application we have in mind.

13.17   **a** $S_{xy} = 44194 - \left[\dfrac{(50)(16705)}{20}\right] = 2431.5$

$S_{xx} = 150 - \left[\dfrac{(50)^2}{20}\right] = 25$

$b = \dfrac{2431.5}{25} = 97.26,$ $a = 835.25 - (97.26)(2.5) = 592.1$

b $\hat{y} = 592.1 + 97.26(2) = 786.62.$ The corresponding residual is $(y - \hat{y}) = 757 - 786.62 = -29.62.$

c   SSResid $= 14194231 - 592.1(16705) - 97.26(44194) = 4892.06$

$s_e = \sqrt{\dfrac{4892.06}{18}} = \sqrt{271.781} = 16.4858$

$s_b = \dfrac{16.4858}{\sqrt{25}} = 3.2972$

The 99% confidence interval for β, the true average change in oxygen usage associated with a one-minute increase in exercise time is

$97.26 \pm (2.88)(3.2972) \Rightarrow 97.26 \pm 9.50 \Rightarrow (87.76, 106.76).$

13.19   **a** $H_o: \beta = 0$ $H_a: \beta \neq 0$

$\alpha = 0.05$ (for illustration)

$t = \dfrac{b}{s_b}$ with df $= 42$

$t = \dfrac{15}{5.3} = 2.8302$

P-value = 2(area under the 42 df t curve to the right of 2.83) $\approx 2(0.0036) = 0.0072.$

Since the P-value is less than α, H_o is rejected. The data supports the conclusion that the simple linear regression model specifies a useful relationship between x and y. (It is advisable to examine a scatter plot of y versus x to confirm the appropriateness of a straight line model for these data).

b $b \pm$ (t critical) $s_b \Rightarrow 15 \pm (2.02)(5.3) \Rightarrow 15 \pm 10.706 \Rightarrow (4.294, 25.706)$

Based on this interval, we estimate the change in mean average SAT score associated with an increase of $1000 in expenditures per child is between 4.294 and 25.706.

13.21 Summary values are: $n = 10$, $\Sigma x = 6{,}970$ $\Sigma x^2 = 5{,}693{,}950$ $\Sigma y = 10{,}148$ $\Sigma y^2 = 12{,}446{,}748$ $\Sigma \ xy = 8{,}406{,}060$.

We first calculate various quantities needed to answer the different parts of this problem.

$$b = \frac{8406060 - \left[\dfrac{(6970)(10148)}{10}\right]}{5693950 - \left[\dfrac{(6970)^2}{10}\right]} = \frac{1332904}{835860} = 1.5946498217404828560$$

$$a = \frac{[1014.8 - (1.5946498217404828560)(697)]}{10} = -96.670925753116550619$$

SSResid $= 12446748 - (-96.670925753116550619)(10148)$
               $- (1.5946498217404828560)(8406060) = 23042.474$

NOTE: Using the formula $SSResid = \sum y^2 - a \sum y - b \sum xy$ can lead to severe roundoff errors unless many significant digits are carried along for the intermediate calculations. The calculations for this problem are particularly prone to roundoff errors because of the large numbers involved. This is the reason we have given many significant digits for the slope and the intercept estimates. You may want to try doing these calculations with fewer significant digits. You will notice a substantial loss in accuracy in the final answer. The formula $SSResid = \sum (y - \hat{y})^2$ provides a more numerically stable alternative. We give the calculations based on this alternative formula in the table below.

We used $b = 1.59465$ and $a = \dfrac{[1014.8 - (1.59465)(697)]}{10} = -96.6711$ to calculate $\hat{y} = a + bx$.

y	$\hat{y} = a + bx$	$y - \hat{y}$	$(y - \hat{y})^2$
303	301.99	1.009	1
491	477.4	13.597	184.9
659	660.79	-1.788	3.2
683	740.52	-57.52	3308.6
922	876.07	45.935	2110
1044	1083.37	-39.37	1550
1421	1306.62	114.379	13082.6
1329	1370.41	-41.407	1714.5
1481	1513.93	-32.925	1084.1

The sum of the numbers in the last column gives $SSResid = 23042.5$ which is accurate to the first decimal place.

$$s_e^2 = \frac{23042.5}{8} = 2880.31$$

$$s_b^2 = \frac{2880.31}{835860} = 0.00344592, \quad s_b = 0.0587020$$

a   The prediction equation is $CHI = -96.6711 + 1.59465 \ Control$.

Using this equation we can predict the mean response time for those suffering a closed-head injury using the mean response time on the same task for individuals with no head injury.

b   Let β denote the expected increase in mean response time for those suffering a closed-head injury associated with a one unit increase in mean response time for the same task for individuals with no head injury.

$H_o: \beta = 0 \qquad H_a: \beta \neq 0$

$\alpha = 0.05$

$$t = \frac{b}{s_b} \qquad \text{with df} = 8$$

$$t = \frac{1.59465}{0.0587020} = 27.1652$$

P-value = 2(area under the 8 df t curve to the right of 27.1652) $\approx$ 2(0) = 0.

259

Since the P-value is less than α, H_o is rejected. The simple linear regression model does provide useful information for predicting mean response times for individuals with CHI and mean response times for the same task for individuals with no head injury. (A scatter plot of y versus x confirms that a straight line model is a reasonable model for this problem).

c   The equation CHI = 1.48 Control says that the mean response time for individuals with CHI is *proportional* to the mean response time for the same task for individuals with no head injury, and the proportionality constant is 1.48. This implies that the mean response time for individuals with CHI is *1.48 times* the mean response time for the same task for individuals with no head injury.

13.23 a   Let β denote the expected change in sales revenue associated with a one unit increase in advertising expenditure.

$H_o: \beta = 0 \qquad H_a: \beta \neq 0$

$\alpha = 0.05$

$$t = \frac{b}{s_b} \qquad \text{with df} = 13$$

$$t = \frac{52.57}{8.05} = 6.53$$

P-value = 2(area under the 13 df t curve to the right of 6.53) $\approx$ 2(0) = 0.

Since the P-value is less than α, H_o is rejected. The simple linear regression model does provide useful information for predicting sales revenue from advertising expenditures.

b $H_o: \beta = 40 \qquad H_a: \beta > 40$

$\alpha = 0.01$

Test statistic: $t = \dfrac{b - 40}{s_b}$ with df = 13

$$t = \frac{(52.57 - 40)}{8.05} = 1.56$$

P-value = area under the 13 df t curve to the right of $1.56 \approx 0.071$.

Since the P-value exceeds α, the null hypothesis is not rejected. The data are consistent with the hypothesis that the change in sales revenue associated with a one unit increase in advertising expenditure does not exceed 40 thousand dollars.

13.25   Let β denote the average change in milk pH associated with a one unit increase in temperature.

$H_o: \beta = 0$ $H_a: \beta < 0$

$\alpha = 0.01$

The test statistic is: $t = \dfrac{b}{s_b}$ with d.f. = 14.

Computations: $n = 16$, $\sum x = 678$, $\sum y = 104.54$,

$$S_{xy} = 4376.36 - \frac{(678)(104.54)}{16} = -53.5225$$

$$S_{xx} = 36056 - \frac{(678)^2}{16} = 7325.75$$

$$b = \frac{-53.5225}{7325.75} = -0.0073$$

$a = 6.53375 - (-0.0073)(42.375) = 6.8431$

SSResid $= 683.447 - 6.8431(104.54) - (-0.00730608)(4376.36) = 0.0177354$

$$s_e = \sqrt{\frac{.0177354}{14}} = \sqrt{.001267} = .0356$$

$$s_b = \frac{.0356}{\sqrt{7325.75}} = .000416$$

$$t = \frac{-0.00730608}{0.000416} = -17.5627$$

P-value = area under the 14 df t curve to the left of $-17.5627 \approx 0$.

Since the P-value is less than α, H₀ is rejected. There is sufficient evidence in the sample to conclude that there is a negative (inverse) linear relationship between temperature and pH. A scatter plot of y versus x confirms this.

Exercises 13.27 – 13.33

13.27

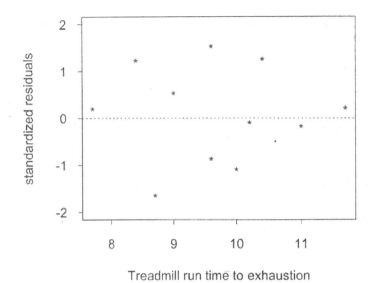

The standardized residual plot does not exhibit any unusual features.

13.29 a The assumptions required in order that the simple linear regression model be appropriate are:

(i) The distribution of the random deviation e at any particular x value has mean value 0.

(ii) The standard deviation of e is the same for any particular value of x.

(iii) The distribution of e at any particular x value is normal.

(iv) The mean value of vigor is a linear function of stem density.

(v) The random deviations $e_1, e_2, \cdots, e_n$ associated with different observations are independent of one another.

b

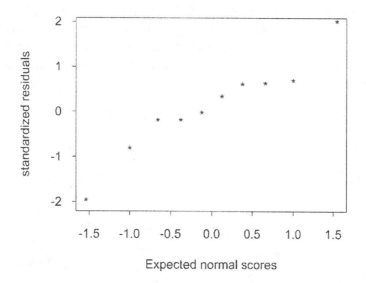

The normal probability plot appears to follow a straight line pattern (approximately). Hence the assumption that the random deviation distribution is normal is plausible.

c

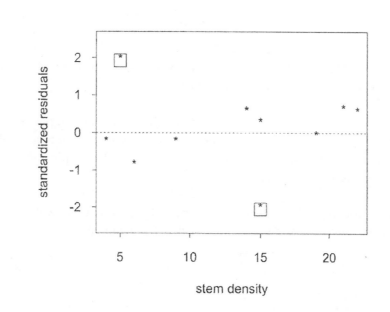

There are two residuals that are relatively large. The corresponding points are enclosed in boxes on the graph above.

d   The negative residuals appear to be associated with small x values, and the positive residuals appear to be associated with large x values. Such a pattern is apparently the result of the fitted regression line being influenced by the two potential outlying points. This would cause one to question the appropriateness of using a simple linear regression model without addressing the issue of outliers.

13.31   **a**

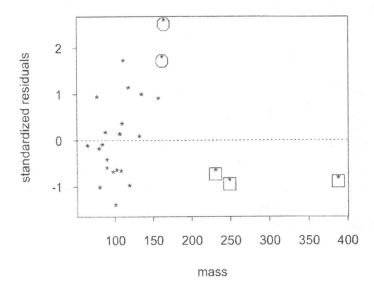

The several large residuals are marked by circles. The potentially influential observations are marked by rectangles.

b   The residuals associated with the potentially influential observations are all negative. Without these three, there appears to be a positive trend to the standardized residual plot. The plot suggests that the simple linear regression model might not be appropriate.

c   There does not appear to be any pattern in the plot that would suggest that it is unreasonable to assume that the variance of y is the same at each x value.

13.33

Year	X	Y	Y-Pred	Residual
1963	188.5	2.26	1.750	0.51000
1964	191.3	2.60	2.478	0.12200
1965	193.8	2.78	3.128	−0.34800
1966	195.9	3.24	3.674	−0.43400
1967	197.9	3.80	4.194	−0.39400
1968	199.9	4.47	4.714	−0.24400
1969	201.9	4.99	5.234	−0.24400
1970	203.2	5.57	5.572	−0.00200
1971	206.3	6.00	6.378	−0.37800
1972	208.2	5.89	6.872	−0.98200
1973	209.9	8.64	7.314	1.32600

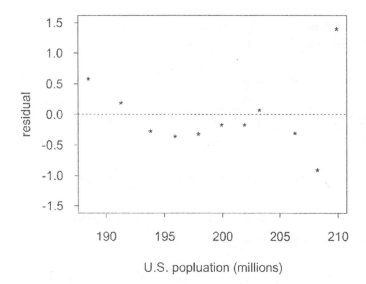

U.S. popluation (millions)

The residuals are positive in 1963 and 1964, then they are negative from 1965 through 1972, followed by a positive residual in 1973. The residuals exhibit a pattern in the plot and thus the plot casts doubt on the appropriateness of the simple linear regression model.

Exercises 13.34 – 13.48

13.35 If the request is for a confidence interval for β the wording would likely be "estimate the <u>change</u> in the average y value associated with a one unit increase in the x variable." If the request is for a confidence interval for $\alpha + \beta x^*$ the wording would likely be "estimate the <u>average y</u> value when the value of the x variable is x^*."

13.37 a $S_{xx} = 2939 - \dfrac{(131)^2}{12} = 2939 - 1430.0833 = 1508.9167$

$s_b = \dfrac{2.33}{\sqrt{1508.9167}} = \dfrac{2.33}{38.8448} = 0.059982$

$b \pm (t \text{ critical})s_b \Rightarrow -0.345043 \pm (2.23)(0.059982)$
$\Rightarrow -0.345043 \pm 0.133761 \Rightarrow (-0.4788, -0.2113)$

b $a + b(1) = 20.125053 - 0.345043(1) = 19.78$

$S_{a+b(1)} = 2.33\sqrt{\dfrac{1}{12} + \dfrac{(1-10.9167)^2}{1508.9167}} = 2.33(0.385365) = 0.8979$

The 95% confidence interval is $19.78 \pm 2.23(0.8979)$
$\Rightarrow 19.78 \pm 2.0023 \Rightarrow (17.78, 21.78)$.

13.39 a The 95% prediction interval for an observation to be made when $x^* = 40$ would be

$6.5511 \pm 2.15 \sqrt{(0.0356)^2 + (0.008955)^2} \Rightarrow 6.5511 \pm 2.15(0.0367)$
$\Rightarrow 6.5511 \pm 0.0789 = (6.4722, 6.6300)$.

b The 99% prediction interval for an observation to be made when $x^* = 35$ would be

$6.5876 \pm 2.98 \sqrt{(0.0356)^2 + (0.009414)^2} \Rightarrow 6.5876 \pm 2.98(0.0368)$
$\Rightarrow 6.5876 \pm 0.1097 = (6.4779, 6.6973)$.

c Yes, because $x^* = 60$ is farther from the mean value of x, which is 42.375, than is 40 or 35.

13.41 a From the MINITAB output we get

Clutch Size = -133.02 + 5.92 Snout-Vent Length

b From the MINITAB output we get $s_b = 1.127$

c Let β denote the mean increase in Clutch size associated with a one unit increase in Snout-Vent length.

$H_0: \beta = 0$ $H_a: \beta > 0$

$\alpha = 0.05$ (a significance level is not specified in the problem so we use 0.05 for illustration)

The test statistic is: $t = \dfrac{b}{s_b}$ with df. = 12.

From the MINITAB output $t = \dfrac{5.919}{1.127} = 5.25$

P-value = area under the 12 df t curve to the right of 5.25 $\approx$ 0.

Hence the null hypothesis is rejected. The data provide strong evidence indicating that the slope is positive.

d The predicted value of the clutch size for a salamander with snout-vent length of 65 is
$-133.02 + 5.919 (65) = 251.715.$

e The value 205 is very much outside the range of snout-vent length values in the available data. The validity of the estimated regression line this far away from the range of x values in the data set is highly questionable. Therefore, calculation of a predicted value and/or a prediction interval for the clutch size for a salamander with snout-vent length of 205 based on available data is not recommended.

13.43 a

$$b = \dfrac{1081.5 - \left[\dfrac{(269)(51)}{14}\right]}{7445 - \left[\dfrac{(269)^2}{14}\right]} = \dfrac{101.571}{2276.357} = 0.04462$$

$a = 3.6429 - (0.04462)(19.214) = 2.78551$

The equation of the estimated regression line is $\hat{y} = 2.78551 + 0.04462x$.

b $H_o: \beta = 0$ $H_a: \beta \neq 0$

$\alpha = 0.05$

The test statistic is: $t = \dfrac{b}{s_b}$ with df. = 12.

$$\text{SSResid} = 190.78 - (2.78551)(51) - (0.00462)(1081.5) = 0.46246$$

$$s_e^2 = \frac{0.46246}{12} = 0.0385$$

$$s_b^2 = \frac{0.0385}{2276.357} = 0.0000169, \quad s_b = 0.004113$$

$$t = \frac{0.04462}{0.004113} = 10.85$$

P-value = 2(area under the 12 df t curve to the right of 10.85) = 2(0) = 0.

Since the P-value is less than α, the null hypothesis is rejected. The data suggests that the simple linear regression model provides useful information for predicting moisture content from knowledge of time.

c   The point estimate of the moisture content of an individual box that has been on the shelf 30 days is $2.78551 + 0.04462(30) = 4.124$.

The 95% prediction interval is

$$4.124 \pm (2.18)\sqrt{0.0385}\sqrt{1 + \frac{1}{14} + \frac{(30 - 19.214)^2}{2276.357}}$$

$$\Rightarrow 4.124 \pm 2.18(0.2079) \Rightarrow 4.124 \pm 0.453 = (3.671, 4.577).$$

d   Since values greater than equal to 4.1 are included in the interval constructed in c, it is very plausible that a box of cereal that has been on the shelf 30 days will not be acceptable.

13.45   a

$$b = \frac{57760 - \left[\dfrac{(1350)(600)}{15}\right]}{155400 - \left[\dfrac{(1350)^2}{15}\right]} = \frac{3760}{33900} = 0.1109$$

$$a = 40 - (0.1109)(90) = 30.019$$

The equation for the estimated regression line is $\hat{y} = 30.019 + 0.1109x$.

b   When $x = 100$, the point estimate of $\alpha + \beta(100)$ is $30.019 + 0.1109(100) = 41.109$.

$$SSResid = 24869.33 - (30.019)(600) - (0.1109)(57760) = 452.346$$

$$s_e^2 = \frac{452.346}{13} = 34.7958$$

$$s_{a+b(100)}^2 = 34.7958\left[\frac{1}{15} + \frac{(100-90)^2}{33900}\right] = 2.422$$

$$s_{a+b(100)} = \sqrt{2.422} = 1.5564$$

The 90% confidence interval for the mean blood level for people who work where the air lead level is 100 is $41.109 \pm (1.77)(1.5564) \Rightarrow 41.109 \pm 2.755 \Rightarrow$ (38.354, 43.864).

c   The prediction interval is $41.109 \pm (1.77)\sqrt{34.7958 + 2.422} \Rightarrow 41.109 \pm 10.798 \Rightarrow (30.311, 51.907)$.

d   The interval of part **b** is for the mean blood level of all people who work where the air lead level is 100. The interval of part **c** is for a single randomly selected individual who works where the air lead level is 100.

13.47 **a**   The 95% prediction interval for sunburn index when distance is 35 is

$$2.5225 \pm 2.16\sqrt{(0.25465)^2 + (0.07026)^2} \Rightarrow 2.5225 \pm 0.5706 \Rightarrow (1.9519, 3.0931).$$

The 95% prediction interval for sunburn index when distance is 45 is

$$1.9575 \pm 2.16\sqrt{(0.25465)^2 + (0.06857)^2} \Rightarrow 1.9575 \pm 0.5696 \Rightarrow (1.3879, 2.5271).$$

The pair of intervals form a set of simultaneous prediction intervals with prediction level of at least $[100 - 2(5)]\% = 90\%$.

b   The simultaneous prediction level would be at least $[100 - 3(1)]\% = 97\%$.

Exercises 13.49 – 13.57

13.49   The quantity r is a statistic as its value is calculated from the sample. It is a measure of how strongly the sample x and y values are linearly related. The value of r is an estimate of ρ. The quantity ρ is a population characteristic. It measures the strength of association between the x and y values in the population.

13.51   Let ρ denote the true correlation coefficient between teaching evaluation index and annual raise.

$H_o: \rho = 0 \qquad H_a: \rho \neq 0$

$\alpha = 0.05$

$$t = \frac{r}{\sqrt{\dfrac{1 - r^2}{n - 2}}} \qquad \text{with df.} = 351$$

$n = 353, \ r = 0.11$

$$t = \frac{0.11}{\sqrt{\dfrac{1 - (0.11)^2}{351}}} = \frac{0.11}{0.05305} = 2.07$$

The t curve with 351 df is essentially the z curve.

P-value = 2(area under the z curve to the right of 2.07) = 2(0.0192) = 0.0384.

Since the P-value is less than α, H_o is rejected. There is sufficient evidence in the sample to conclude that there appears to be a linear association between teaching evaluation index and annual raise.

According to the guidelines given in the text book, r = 0.11 suggests only a weak linear relationship. Since $r^2 = 0.0121$, fitting the simple linear regression model to the data would result in only about 1.21% of observed variation in annual raise being explained.

13.53   **a**   Let ρ denote the correlation coefficient between time spent watching television and grade point average in the population from which the observations were selected.

$H_o: \rho = 0 \qquad H_a: \rho < 0$

$\alpha = 0.01$

$$t = \frac{r}{\sqrt{\dfrac{1-r^2}{n-2}}} \quad \text{with df.} = 526$$

$$n = 528, \ r = -0.26$$

$$t = \frac{-0.26}{\sqrt{\dfrac{1-(-0.26)^2}{526}}} = \frac{-0.26}{0.042103} = -6.175$$

The t curve with 526 df is essentially the z curve.

P-value = area under the z curve to the left $-6.175 \approx 0$.

Since the P-value is less than α, H₀ is rejected. The data does support the conclusion that there is a negative correlation in the population between the two variables, time spent watching television and grade point average.

b   The coefficient of determination measures the proportion of observed variation in grade point average explained by the regression on time spent watching television. This value would be $(-0.26)^2 = 0.0676$. Thus only 6.76% of the observed variation in grade point average would be explained by the regression. This is not a substantial percentage.

13.55   From the summary quantities:

$$S_{xy} = 673.65 - \left[\frac{(136.02)(39.35)}{9} \right] = 78.94$$

$$S_{xx} = 3602.65 - \left[\frac{(136.02)^2}{9} \right] = 1546.93$$

$$S_{yy} = 184.27 - \left[\frac{(39.35)^2}{9} \right] = 12.223$$

$$r = \frac{78.94}{\sqrt{(1546.93)(12.223)}} = \frac{78.94}{137.51} = 0.574$$

Let ρ denote the correlation between surface and subsurface concentration.

$H_o: \rho = 0$ $H_a: \rho \neq 0$

$\alpha = 0.05$

$$t = \frac{r}{\sqrt{\dfrac{1 - r^2}{n - 2}}} \quad \text{with df.} = 7$$

$$t = \frac{0.574}{\sqrt{\dfrac{1 - (0.574)^2}{7}}} = 1.855$$

P-value = 2(area under the 7 df t curve to the right of 1.855) $\approx$ 2(0.053) = 0.106.

Since the P-value exceeds α, H_o is not rejected. The data does not support the conclusion that there is a linear relationship between surface and subsurface concentration.

13.57 $H_o: \rho = 0$ $H_a: \rho \neq 0$

$\alpha = 0.05$

$$t = \frac{r}{\sqrt{\dfrac{1 - r^2}{n - 2}}} \quad \text{with df.} = 9998$$

n = 10000, r = 0.022

$$t = \frac{0.022}{\sqrt{\dfrac{1 - (0.022)^2}{9998}}} = 2.2$$

The t curve with 9998 df is essentially the z curve.

P-value = 2(area under the z curve to the right of 2.2) = 2(0.0139) = 0.0278.

Since the P-value is less than α, H_o is rejected. The results are statistically significant. Because of the extremely large sample size, it is easy to detect a value of ρ which differs from zero by a small amount. If ρ is very close to zero, but not zero, the practical significance of a non-zero correlation may be of little consequence.

Exercises 13.58 – 13.76

13.59   **a** $$t = \frac{-0.18}{\sqrt{\dfrac{1-(-.18)^2}{345}}} = \frac{-0.18}{0.052959} = -3.40 \qquad \text{with df} = 345$$

If the test was a one-sided test, then the P-value equals the area under the z curve to the left of –3.40, which is equal to 0.0003. It the test was a two-sided test, then the P-value is 2(0.0003) = 0.0006. While the researchers' statement is true, they could have been more precise in their statement about the P-value.

b   From my limited experience, I have observed that the more visible a person's sense of humor, the less depressed they *appear* to be. This would suggest a negative correlation between Coping Humor Scale and Sense of Humor.

c   Since $r^2 = (-0.18)^2 = 0.0324$, only about 3.24% of the observed variability in sense of humor can be explained by the linear regression model. This suggests that a simple linear regression model may not give accurate predictions.

13.61   **a** $H_0: \beta = 0 \qquad H_a: \beta \neq 0$

$\alpha = 0.01$

$$t = \frac{b}{s_b}$$

From the Minitab output, t = –3.95 and the P-value = 0.003. Since the P-value is less than α, H_0 is rejected. The data supports the conclusion that the simple linear regression model is useful.

b   A 95% confidence interval for β is $-2.3335 \pm 2.26(0.5911) \Rightarrow -2.3335 \pm 1.3359$ $\Rightarrow (-3.6694, -0.9976)$.

c   a+b(10) = 88.796 – 2.3335(10) = 65.461

$s_{a+b(10)} = 0.689$

The 95% prediction interval for an individual y when x = 10 is

$$65.461 \pm 2.26\sqrt{(0.689)^2 + 4.789} \Rightarrow 65.461 \pm 5.185 \Rightarrow (60.276, 70.646).$$

d   Because x = 11 is farther from $\bar{x}$ than x = 10 is from $\bar{x}$.

13.63 **a**   Let ρ denote the correlation coefficient between soil hardness and trail length.

$H_0: \rho = 0 \quad H_a: \rho < 0$

$\alpha = 0.05$

$$t = \frac{r}{\sqrt{\dfrac{1-r^2}{(n-2)}}} \quad \text{with df.} = 59$$

$$t = \frac{-0.6213}{\sqrt{\dfrac{1-(-0.6213)^2}{59}}} = -6.09$$

P-value = area under the 59 df t curve to the left of $-6.09 \approx 0$.

Since the P-value is less than α, the null hypothesis is rejected. The data supports the conclusion of a negative correlation between trail length and soil hardness.

b   When $x^* = 6$, $a+b(6) = 11.607 - 1.4187(6) = 3.0948$

$$s^2_{a+b(6)} = (2.35)^2\left[\frac{1}{61} + \frac{(6-4.5)^2}{250}\right] = 0.1402$$

$$s_{a+b(6)} = \sqrt{0.1402} = 0.3744$$

The 95% confidence interval for the mean trail length when soil hardness is 6 is

$3.0948 \pm 2.00(0.3744) \Rightarrow 3.0948 \pm 0.7488 \Rightarrow (2.346, 3.844)$.

c   When $x^* = 10$, $a+b(10) = 11.607 - 1.4187(10) = -2.58$

According to the least-squares line, the predicted trail length when soil hardness is 10 is -2.58. Since trail length cannot be negative, the predicted value makes no sense. Therefore, one would not use the simple linear regression model to predict trail length when hardness is 10.

13.65 $n = 17$, $\Sigma x = 821$, $\Sigma x^2 = 43447$, $\Sigma y = 873$, $\Sigma y^2 = 46273$, $\Sigma xy = 40465$,

$$S_{xy} = 40465 - \left[\frac{(821)(873)}{17}\right] = 40465 - 42160.7647 = -1695.7647$$

$$S_{xx} = 43447 - \left[\frac{(821)^2}{17}\right] = 43447 - 39649.4706 = 3797.5294$$

$$S_{yy} = 46273 - \left[\frac{(873)^2}{17}\right] = 46273 - 44831.1176 = 1441.8824$$

$$b = \frac{-1695.7647}{3797.5294} = -0.4465$$

$$a = 51.3529 - (-0.4465)(48.2941) = 72.9162$$

$$SSResid = 46273 - 72.9162(873) - (-0.4465)(40465) = 684.78$$

$$s_e^2 = \frac{684.78}{15} = 45.652, \quad s_e = 6.7566$$

$$s_b = \frac{6.7566}{\sqrt{3797.5294}} = 0.1096$$

a   Let β denote the average change in percentage area associated with a one year increase in age.

   $H_o: \beta = -0.5$ $H_a: \beta \neq -0.5$

   $\alpha = 0.10$

   $$t = \frac{b - (-0.5)}{s_b} \quad \text{with df.} = 15$$

   $$t = \frac{-0.4465 - (-0.5)}{0.1096} = 0.49$$

   P-value = 2(area under the 15 df t curve to the right of 0.49) $\approx$ 2(0.312) = 0.624.

   Since the P-value exceeds α, H_o is not rejected. There is not sufficient evidence in the sample to contradict the prior belief of the researchers.

b   When $x^* = 50$, $\hat{y} = 72.9162 + (-0.4465)(50) = 50.591$

$$S_{a+b(50)} = 6.7471 \sqrt{\frac{1}{17} + \frac{(50 - 48.2941)^2}{3797.5294}} = 1.647$$

The 95% confidence interval for the true average percent area covered by pores for all 50 year-olds is

$$50.591 \pm (2.13)(1.647) \Rightarrow 50.591 \pm 3.508 \Rightarrow (47.083, 54.099).$$

13.67 The 95% confidence interval for α is

$$20.1251 \pm 2.23(0.9402) \Rightarrow 20.1251 \pm 2.0967 \Rightarrow (18.0284, 22.2218).$$

13.69 For data set 1:

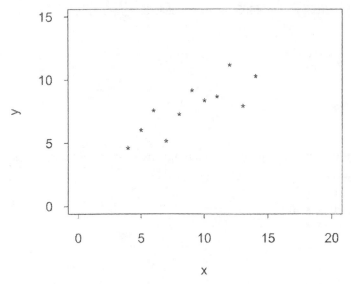

The plot above supports the appropriateness of fitting a simple linear regression model to data set 1.

For data set 2:

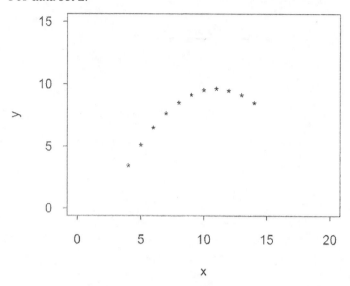

This plot suggests quite clearly that the fitting of a simple linear regression model to data set 2 would not be appropriate.

For data set 3:

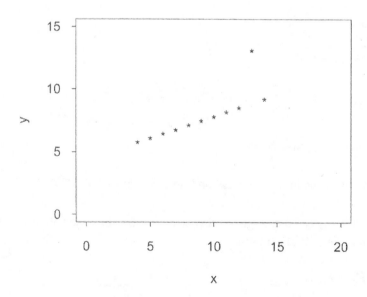

This plot reveals an observation which would have an unusually large residual. A simple linear regression model would not be appropriate for this data set, but might be for the data set with the one unusual observation deleted.

For data set 4:

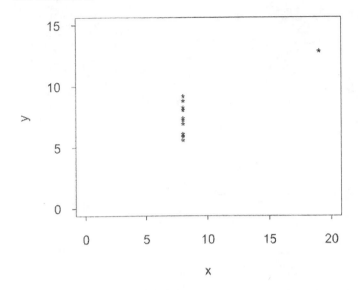

This plot reveals one point that is a very influential point. In fact, the slope is determined solely by this observation. The simple linear regression model would not be appropriate for this data set.

13.71 Summary values are: n = 8,

$S_{xx} = 42$, $S_{yy} = 586.875$, $S_{xy} = 1.5$

a $b = \dfrac{1.5}{42} = 0.0357$

$a = 58.125 - (0.0357)(4.5) = 57.964$

The equation of the estimated regression line is $\hat{y} = 57.964 + 0.0357x$.

b   Let β denote the expected change in glucose concentration associated with a one day increase in fermentation time.

$H_o: \beta = 0$ $H_a: \beta \neq 0$

$\alpha = 0.10$

The test statistic: $t = \dfrac{b}{s_b}$ with df. = 6.

278

From the data, $s_b = 1.526$ and $t = \dfrac{0.0357}{1.526} = 0.023$.

P-value = 2(area under the 6 df t curve to the right of 0.023) = 0.982.

Since the P-value exceeds α, the null hypothesis is not rejected. The data does not indicate a linear relationship between fermentation time and glucose concentration.

c

x	y	Pred-y	Residual
1	74	58.00	16.00
2	54	58.04	−4.04
3	52	58.07	−6.07
4	51	58.11	−7.11
5	52	58.14	−6.14
6	53	58.18	−5.18
7	58	58.21	−0.21
8	71	58.25	12.75

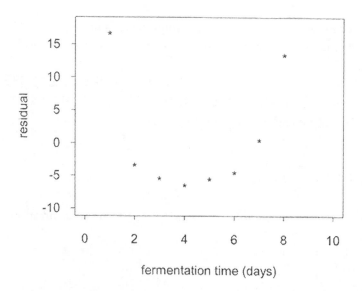

d   The residual plot has a very distinct curvilinear pattern which indicates that a simple linear regression model is not appropriate for describing the relationship between y and x. Instead, a model incorporating curvature should be fit.

13.73 **a** Summary values are: n = 17, S_{xx} = 13259.0706, S_{yy} = 1766.4706, S_{xy} = −79.9294, where x = depth and y = zinc concentration.

$$r = \frac{-79.9294}{\sqrt{(13259.0706)(1766.4706)}} = -0.0165$$

Let ρ denote the correlation between depth and zinc concentration.

H_o: $\rho = 0$ H_a: $\rho \neq 0$

$\alpha = 0.05$

$$t = \frac{r}{\sqrt{\frac{1-r^2}{(n-2)}}} \quad \text{with df = 15.}$$

$$t = \frac{(-0.0165)}{\sqrt{\frac{1-(-0.0165)^2}{15}}} = -0.06$$

P-value = 2(area under the 15 df t curve to the left of −0.06) ≈ 2(0.47) = 0.94.

Since the P-value exceeds α, the null hypothesis is not rejected. The data suggests that no correlation exists between depth and zinc concentration.

b Summary values are: n = 17, $\sum x$ = 531.7, $\sum x^2$ = 29,888.77, $\sum y$ = 58.5, $\sum y^2$ = 204.51, $\sum xy$ = 1983.64, S_{xx} = 13259.0706, S_{yy} = 3.2012, S_{xy} = 153.9665, where x = depth and y = iron concentration.

$$r = \frac{153.9665}{\sqrt{(13259.30706)(3.2012)}} = 0.747$$

Let ρ denote the correlation between depth and iron concentration.

H_o: $\rho = 0$ H_a: $\rho \neq 0$

$\alpha = 0.05$

$$t = \frac{r}{\sqrt{\frac{1-r^2}{(n-2)}}} \quad \text{with df. = 15}$$

$$t = \frac{0.747}{\sqrt{\dfrac{1-(0.747)^2}{15}}} = 4.35$$

P-value = 2(area under the 15 df t curve to the right of 4.35) $\approx$ 2(0.0003) = 0.0006.

Since the P-value is less than α, the null hypothesis is rejected. The data does suggest a correlation between depth and iron concentration.

c $\qquad b = \dfrac{153.9665}{13259.0706} = 0.01161$

$\qquad a = \dfrac{58.5 - (0.01161)(531.7)}{17} = 3.0781$

d $\qquad$ When $x^* = 50$, $a+b(50) = 3.6586$.

$\qquad$ SSResid = $204.51 - (3.0781)(58.5) - (0.01161)(1983.64) = 1.411$

$\qquad s_e^2 = \dfrac{1.411}{15} = 0.0941$

$\qquad s_{a+b(50)}^2 = 0.0941\left[\dfrac{1}{17} + \dfrac{(50-31.276)^2}{13259.0706}\right] = 0.00802$

$\qquad$ The 95% prediction interval for the iron concentration of a single core sample taken at a depth of 50m. is

$\qquad 3.6586 \pm (2.13)\sqrt{0.0941 + 0.00802} \Rightarrow 3.6586 \pm (2.13)(0.3196)$
$\qquad \Rightarrow 3.6585 \pm 0.6807 \Rightarrow (2.9778, 4.3392)$.

e $\qquad$ When $x^* = 70$, $a+b(70) = 3.8908$

$\qquad s_{a+b(70)}^2 = 0.0941\left[\dfrac{1}{17} + \dfrac{(70-31.276)^2}{13259.0706}\right] = 0.0162$

$\qquad s_{a+b(70)} = \sqrt{0.0162} = 0.1273$

$\qquad$ The 95% confidence interval for $\alpha + \beta(70)$ is

$\qquad 3.8908 \pm (2.13)(0.1273) \Rightarrow 3.8908 \pm 0.2711 \Rightarrow (3.6197, 4.1619)$.

With 95% confidence it is estimated that the mean iron concentration at a depth of 70m is between 3.6197 and 4.1619.

13.75 a The e_i's are the deviations of the observations from the population regression line, whereas the residuals are the deviations of the observations from the estimated regression line.

b The simple linear regression model states that $y = \alpha + \beta x + e$. Without the random deviation e, the equation implies a deterministic model, whereas the simple linear regression model is probabilistic.

c The quantity b is a statistic. Its value is known once the sample has been collected, and different samples result in different b values. Therefore, it does not make sense to test hypotheses about b. Only hypotheses about a population characteristic can be tested.

d If r = +1 or –1, then each point falls exactly on the regression line and SSResid would equal zero. A true statement is that SSResid is always greater than or equal to zero.

e The sum of the residuals must equal zero. Thus, if they are not all exactly zero, at least one must be positive and at least one must be negative. They cannot all be positive. Since there are some positive and no negative values among the reported residuals, the student must have made an error.

f SSTo = $\sum(y-\bar{y})^2$ must be greater than or equal to SSResid = $\sum(y-\hat{y})^2$. Thus, the values given must be incorrect.

Chapter 14

Exercises 14.1 – 14.14

14.1   A deterministic model does not have the random deviation component e, while a probabilistic
model does contain such a component.

Let $y =$ total number of goods purchased at a service station which sells only one
grade of gas and one type of motor oil.

   $x_1 =$ gallons of gasoline purchased

   $x_2 =$ number of quarts of motor oil purchased.

Then y is related to x_1 and x_2 in a deterministic fashion.

Let $y =$ IQ of a child

   $x_1 =$ age of the child

   $x_2 =$ total years of education of the parents.

Then y is related to x_1 and x_2 in a probabilistic fashion.

14.3   The following multiple regression model is suggested by the given statement.
$$y = \beta_0 + \beta_1 x_1 + \beta_2 x_2 + e.$$

An interaction term is not included in the model because it is given that x_1 and x_2 make
independent contributions to academic achievement.

14.5   **a** $415.11 - 6.6(20) - 4.5(40) = 103.11$

   b $415.11 - 6.6(18.9) - 4.5(43) = 96.87$

   c $\beta_1 = -6.6$. Hence 6.6 is the expected *decrease* in yield associated with a one-unit
increase in mean temperature between date of coming into hop and date of
picking when the mean percentage of sunshine remains fixed.

   $\beta_2 = -4.5$. So 4.5 is the expected *decrease* in yield associated with a one-unit
increase in mean percentage of sunshine when mean temperature remains fixed.

14.7   a

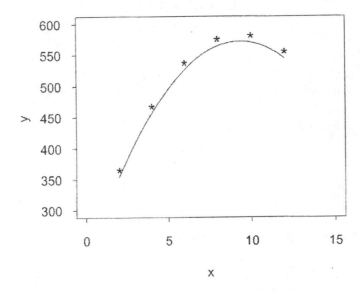

b   The mean chlorine content at x = 8 is 564, while at x = 10 it is 570. So the mean chlorine content is higher for x = 10 than for x = 8.

c   When x = 9, y = 220 + 75(9) − 4(9)² = 571.

The change in mean chlorine content when the degree of delignification increases from 8 to 9 is 571 − 564 = 7.

The change in mean chlorine content when the degree of delignification increases from 9 to 10 is 570 − 571 = −1.

14.9   **a**

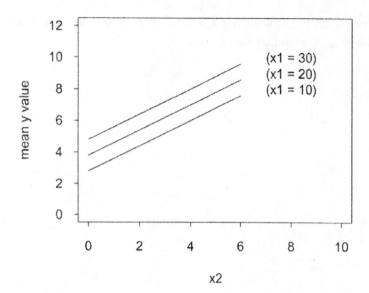

For $x_1 = 30$, $y = 4.8 + 0.8x_2$
For $x_1 = 20$, $y = 3.8 + 0.8x_2$
For $x_1 = 10$, $y = 2.8 + 0.8x_2$

b

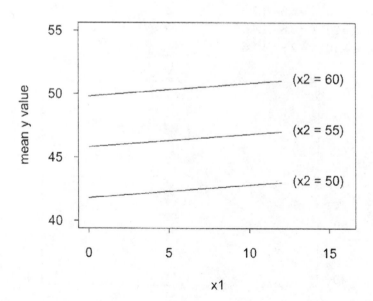

For $x_2 = 60$, $y = 49.8 + 0.1x_2$
For $x_2 = 55$, $y = 45.8 + 0.1x_2$
For $x_2 = 50$, $y = 41.8 + 0.1x_2$

c The parallel lines in each graph are attributable to the lack of interaction between the two independent variables.

d

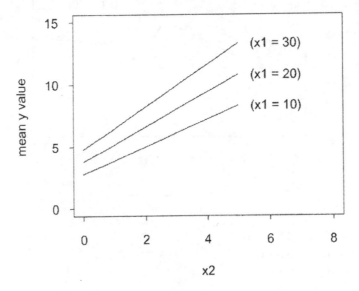

For $x_1 = 30$, $y = 4.8 + 1.7x_2$
For $x_1 = 20$, $y = 3.8 + 1.4x_2$
For $x_1 = 10$, $y = 2.8 + 1.1x_2$

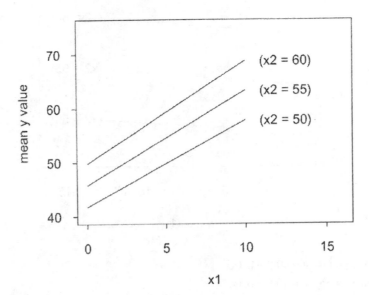

For $x_2 = 60$, $y = 49.8 + 1.9x_2$
For $x_2 = 55$, $y = 45.8 + 1.75x_2$
For $x_2 = 50$, $y = 41.8 + 1.6x_2$

Because there is an interaction term, the lines are not parallel.

14.11 **a** $y = \alpha + \beta_1x_1 + \beta_2x_2 + \beta_3x_3 + e$

    b $y = \alpha + \beta_1x_1 + \beta_2x_2 + \beta_3x_3 + \beta_4x_1^2 + \beta_5x_2^2 + \beta_6x_3^2 + e$

    c $y = \alpha + \beta_1x_1 + \beta_2x_2 + \beta_3x_3 + \beta_4x_1x_2 + e$
       $y = \alpha + \beta_1x_1 + \beta_2x_2 + \beta_3x_3 + \beta_4x_1x_3 + e$
       $y = \alpha + \beta_1x_1 + \beta_2x_2 + \beta_3x_3 + \beta_4x_2x_3 + e$

    d $y = \alpha + \beta_1x_1 + \beta_2x_2 + \beta_3x_3 + \beta_4x_1^2 + \beta_5x_2^2 + \beta_6x_3^2 + \beta_7x_1x_2 + \beta_8x_1x_3 + \beta_9x_2x_3 + e$

14.13 **a** Three dummy variables would be needed to incorporate a non-numerical variable with four categories.

    $x_3 = 1$ if the car is a sub-compact, 0 otherwise
    $x_4 = 1$ if the car is a compact, 0 otherwise
    $x_5 = 1$ if the car is a midsize, 0 otherwise

    $y = \alpha + \beta_1x_1 + \beta_2x_2 + \beta_3x_3 + \beta_4x_4 + \beta_5x_5 + e$

    b $x_6 = x_1x_3, x_7 = x_1x_4,$ and $x_8 = x_1x_5$ are the additional predictors needed to incorporate interaction between age and size class.

Exercises 14.15 – 14.35

14.15 **a** $b_3 = -0.0096$ is the estimated change in mean VO₂ max associated with a one-unit increase in "one mile walk time" when the value of the other predictor variables are fixed. The change is actually a decrease since the regression coefficient is negative.

    b $b_1 = 0.6566$ is the estimated difference in mean VO₂ max for males versus females (male minus female) when the values of the other predictor variables are fixed.

    c $\hat{y}$ $= 3.5959 + .6566(1) + .0096(80) - .0996(11) - .0080(140)$
          $= 3.5959 + .6566 + .7680 - 1.0956 - 1.12$
          $= 2.8049$

$$\text{residual} = y - \hat{y} = 3.15 - 2.8049 = 0.3451$$

d $$R^2 = 1 - \frac{SSResid}{SSTo} = 1 - \frac{30.1033}{102.3922} = 1 - .294 = .706$$

e $$s_e^2 = \frac{SSResid}{n-(k+1)} = \frac{30.1033}{196-5} = \frac{30.1033}{191} = .157609$$

$$s_e = \sqrt{.157609} = .397$$

14.17 **a** $0.05 > \text{P-value} > 0.01$

b $\text{P-value} > 0.10$

c $\text{P-value} = 0.01$

d $0.01 > \text{P-value} > 0.001$

14.19 The fitted model was $y = \alpha + \beta_1 x_1 + \beta_2 x_2 + \beta_3 x_3 + \beta_4 x_4 + \beta_5 x_5 + \beta_6 x_6 + e$

H_o: $\beta_1 = \beta_2 = \beta_3 = \beta_4 = \beta_5 = \beta_6 = 0$
H_a: at least one among β_1, β_2, β_3, β_4, β_5, β_6 is not zero

$\alpha = 0.01$

$$F = \frac{R^2/k}{(1-R^2)/[n-(k+1)]}$$

$n = 37$, $df_1 = k = 6$, $df_2 = n - (k+1) = 37 - 7 = 30$

$R^2 = 0.83$

$$F = \frac{R^2/k}{(1-R^2)/[n-(k+1)]} = \frac{.83/6}{(1-.83)/30} = 24.41$$

From Appendix Table VII, $0.001 > \text{P-value}$.

Since the P-value is less than α, the null hypothesis is rejected. There does appear to be a useful linear relationship between y and at least one of the six predictors.

14.21   The fitted model was $y = \alpha + \beta_1 x_1 + \beta_2 x_2 + \beta_3 x_3 + \beta_4 x_4 + e$

H_o: $\beta_1 = \beta_2 = \beta_3 = \beta_4 = 0$
H_a: at least one among β_1, β_2, β_3, β_4, is not zero

$\alpha = 0.05$

$$F = \frac{R^2/k}{(1-R^2)/[n-(k+1)]}$$

$n = 196$, $df_1 = k = 4$, $df_2 = n - (k+1) = 196 - 5 = 191$

$R^2 = 0.706$

$$F = \frac{R^2/k}{(1-R^2)/[n-(k+1)]} = \frac{.706/4}{(1-.706)/191} = 114.66$$

From Appendix Table VII, $0.001 > $ P-value.

Since the P-value is less than α, the null hypothesis is rejected. There does appear to be a useful linear relationship between y and at least one of the four predictors.

14.23   **a**   Estimated mean value of $y = 86.85 - 0.12297x_1 + 5.090x_2 - 0.07092x_3 + 0.001538x_4$

   b H_o: $\beta_1 = \beta_2 = \beta_3 = \beta_4 = 0$
                H_a: at least one among β_1, β_2, β_3, β_4, is not zero

$\alpha = 0.01$

$$F = \frac{R^2/k}{(1-R^2)/[n-(k+1)]}$$

$n = 31$, $df_1 = k = 4$, $df_2 = n - (k+1) = 31 - 5 = 26$

$R^2 = 0.908$

$$F = \frac{R^2/k}{(1-R^2)/[n-(k+1)]} = \frac{.908/4}{(1-.908)/26} = 64.15$$

From Appendix Table VII, $0.001 > $ P-value

Since the P-value is less than α, the null hypothesis is rejected. There does appear to be a useful linear relationship between y and at least one of the four predictors.

c $R^2 = 0.908$. This means that 90.8% of the variation in the observed tar content values has been explained by the fitted model.

$s_e = 4.784$. This means that the typical distance of an observation from the corresponding mean value is 4.784.

14.25   **a**   The regression model being fitted is $y = \alpha + \beta_1 x_1 + \beta_2 x_2 + e$.
Using MINITAB, the regression command yields the following output.

The regression equation is
weight(g) = - 511 + 3.06 length(mm) - 1.11 age(years)

Predictor	Coef	SE Coef	T	P
Constant	510.9	286.1	-1.79	0.096
length(mm)	3.0633	0.8254	3.71	0.002
age(years)	1.113	9.040	0.12	0.904

S = 94.24   R-Sq = 59.3%   R-Sq(adj) = 53.5%

Analysis of Variance

Source	DF	SS	MS	F	P
Regression	2	181364	90682	10.21	0.002
Residual Error	14	124331	8881		
Total	1	305695			

b H_o: $\beta_1 = \beta_2 = 0$
H_a: At least one of the two β_i's is not zero.

$$F = \frac{SSRegr / k}{SSResid / [n - (k + 1)]}$$

n = 17, $df_1 = k = 2$, $df_2 = n - (k + 1) = 17 - 3 = 14$

$$F = \frac{SSRegr / k}{SSResid / [n - (k + 1)]} = \frac{181364/2}{124331/14} = 10.21$$

From the Minitab output, the P-value = 0.002. Since the P-value is less than 0.05 (we have chosen $\alpha = 0.05$ for illustration) the null hypothesis is rejected. The data suggests that the multiple regression model is useful for predicting weight.

14.27   **a**   Using MINITAB to fit the required regression model yields the following output.

The regression equation is
volume = - 859 + 23.7 minwidth + 226 maxwidth + 225 elongation

Predictor	Coef	SE Coef	T	P
Constant	-859.2	272.9	-3.15	0.005
minwidth	23.72	85.66	0.28	0.784
maxwidth	225.81	85.76	2.63	0.015
elongation	225.24	90.65	2.48	0.021

S = 287.0   R-Sq = 67.6%   R-Sq(adj) = 63.4%

Analysis of Variance

Source	DF	SS	MS	F	P
Regression	3	3960700	1320233	16.03	0.000
Residual Error	23	1894141	82354		
Total	26	5854841			

   b   Adjusted R^2 takes into account the number of predictors used in the model whereas R^2 does not do so. In particular, adjusted R^2 enables us to make a "fair comparison" of the performances of models with differing numbers of predictors.

   c   We test
$H_o: \beta_1 = \beta_2 = \beta_3 = 0$
H_a: At least one of the three β_i's is not zero.

$\alpha = 0.05$ (for illustration)

$$F = \frac{R^2/k}{(1-R^2)/[n-(k+1)]}$$

n = 27, $df_1 = k = 3$, $df_2 = n - (k + 1) = 27 - 4 = 23$

$R^2 = 0.676$

$$F = \frac{R^2/k}{(1-R^2)/[n-(k+1)]} = \frac{0.676/3}{(1-0.676)/23} = 16.03$$

The corresponding P-value is 0.000 (correct to 3 decimals). Since the P-value is less than α, the null hypothesis is rejected. There does appear to be a useful linear relationship between y and at least one of the three predictors.

14.29   **a**   SSResid = 390.4347

SSTo = $7855.37 - 14(21.1071)^2 = 1618.2093$

SSRegr = $1618.2093 - 390.4347 = 1227.7746$

b $R^2 = \dfrac{1227.7746}{1618.2093} = .759$

This means that 75.9 percent of the variation in the observed shear strength values has been explained by the fitted model.

c H_0: $\beta_1 = \beta_2 = \beta_3 = \beta_4 = \beta_5 = 0$
H_a: at least one among $\beta_1, \beta_2, \beta_3, \beta_4, \beta_5$, is not zero

$\alpha = 0.05$

$$F = \frac{R^2/k}{(1-R^2)/[n-(k+1)]}$$

n = 14, df$_1$ = k = 5, df$_2$ = n – (k + 1) = 14 – 6 = 8

$R^2 = 0.759$

$$F = \frac{R^2/k}{(1-R^2)/[n-(k+1)]} = \frac{.759/5}{(1-.759)/8} = 5.039$$

From Appendix Table VII, 0.05 > P-value >0.01.

Since the P-value is less than α, the null hypothesis is rejected. There does appear to be a useful linear relationship between y and at least one of the predictors. The data suggests that the independent variables as a group do provide information that is useful for predicting shear strength.

14.31 $H_0: \beta_1 = \beta_2 = 0$

$H_a:$ At least one of the two β_i's is not zero.

$\alpha = 0.01$

$$F = \frac{R^2/k}{(1-R^2)/[n-(k+1)]}$$

$n = 24,\ \ df_1 = k = 2,\ \ df_2 = n - (k+1) = 24 - 3 = 21$

$$F = \frac{R^2/k}{(1-R^2)/[n-(k+1)]} = \frac{.902/2}{(1-.902)/21} = 96.64$$

From Appendix Table VII, 0.001 > P-value.

Since the P-value is less than α, the null hypothesis is rejected. The data suggests that the quadratic model does have utility for predicting yield.

14.33   Using MINITAB, the regression command yields the following output.

The regression equation is

Y = – 151 – 16.2 X1 + 13.5 X2 + 0.0935 X1-SQ – 0.253 X2-SQ + 0.0492 X1*X2.

Predictor	Coef	Stdev	t-ratio	p
Constant	−151.4	134.1	−1.13	0.292
X1	−16.216	8.831	−1.84	0.104
X2	13.476	8.187	1.65	0.138
X1-SQ	0.09353	0.07093	1.32	0.224
X2-SQ	−0.2528	0.1271	−1.99	0.082
X1*X2	0.4922	0.2281	2.16	0.063

It can be seen (except for differences due to rounding errors) that the estimated regression equation given in the problem is correct.

14.35 Using MINITAB, the regression command yields the following output.

The regression equation is INF_RATE = 35.8 − 0.68 AVE_TEMP + 1.28 AVE_RH

Predictor	Coef	Stdev	t-ratio	p
Constant	35.83	53.54	0.67	0.508
AVE_TEMP	−0.676	1.436	−0.47	0.641
AVE_RH	1.2811	0.4243	3.02	0.005

s = 22.98   R-sq = 55.0%   R-sq(adj) = 52.1%

Analysis of Variance

SOURCE	DF	SS	MS	F	p
Regression	2	20008	10004	18.95	0.000
Error	31	16369	528		
Total	33	36377			

H_o: $\beta_1 = \beta_2 = 0$
H_a: At least one of the two β_i's is not zero.

$$F = \frac{SSRegr/k}{SSResid/[n-(k+1)]}$$

n = 34, $df_1 = k = 2$, $df_2 = n - (k + 1) = 34 - 3 = 31$

$$F = \frac{SSRegr / k}{SSResid /[n-(k+1)]} = \frac{20008/2}{16369/31} = 18.95$$

From the Minitab output, the P-value ≈ 0. Since the P-value is less than α, the null hypothesis is rejected. The data suggests that the multiple regression model has utility for predicting infestation rate.

Exercises 14.36 – 14.50

14.37 a The degrees of freedom for error is 100 − (7 + 1) = 92. From Appendix Table III, the critical t value is approximately 1.99.

The 95% confidence interval for β_3 is
−0.489 ± (1.99)(0.1044) ⇒ −4.89 ± 0.208 ⇒ (−0.697, −0.281).

With 95% confidence, the change in the mean value of a vacant lot associated with a one unit increase in distance from the city's major east-west thoroughfare is a decrease of as little as 0.281 or as much as 0.697.

b $H_o: \beta_1 = 0$ $H_a: \beta_1 \neq 0$
$\alpha = 0.05$

$t = \dfrac{b_1}{s_{b_1}}$ with df. = 92

$t = \dfrac{-.183}{.3055} = -0.599$

P-value = 2(area under the 92 df t curve to the left of –0.599) ≈ 2(0.275) = 0.550.

Since the P-value exceeds α, the null hypotheses is not rejected. This means that there is not sufficient evidence to conclude that there is a difference in the mean value of vacant lots that are zoned for residential use and those that are not zoned for residential use.

14.39   **a** $H_o: \beta_1 = \beta_2 = 0$
H_a: At least one of the two β_i's is not zero.

$\alpha = 0.05$

$$F = \dfrac{R^2/k}{(1 - R^2)/[n - (k+1)]}$$

n = 50, $df_1 = k = 2$, $df_2 = n - (k + 1) = 50 - 3 = 47$, $R^2 = 0.86$

$$F = \dfrac{R^2/k}{(1 - R^2)/[n - (k+1)]} = \dfrac{.86/2}{(1 - .86)/47} = 144.36$$

From Appendix Table VII, 0.001 > P-value.
Since the P-value is less than α, the null hypothesis is rejected. The data suggests that the quadratic regression model has utility for predicting MDH activity.

b $H_o: \beta_2 = 0$ $H_a: \beta_2 \neq 0$

$\alpha = 0.01$

$t = \dfrac{b_2}{s_{b_2}}$ with df. = 47

$$t = \frac{.0446}{.0103} = 4.33$$

P-value = 2(area under the 47 df t curve to the right of 4.33) $\approx$ 2(0) = 0.

Since the P-value is less than α, the null hypothesis is rejected. The quadratic term is an important term in this model.

c   The point estimate of the mean value of MDH activity for an electrical conductivity level of 40 is $-0.1838 + 0.0272(40) + 0.0446(40^2) = -0.1838 + 0.0272(40) + 0.0446(1600) = 72.2642$.

The 90% confidence interval for the mean value of MDH activity for an electrical conductivity level of 40 is
$$72.2642 \pm (1.68)(0.120) \Rightarrow 72.2642 \pm 0.2016 \Rightarrow (72.0626, 72.4658)$$

14.41 a   The value 0.469 is an estimate of the expected change (increase) in the mean score of students associated with a one unit increase in the student's expected score holding time spent studying and student's grade point average constant.

b H_0: $\beta_1 = \beta_2 = \beta_3 = 0$
H_a: At least one of the three β_i's is not zero.

$\alpha = 0.05$

$$F = \frac{R^2/k}{(1-R^2)/[n-(k+1)]}$$

$n = 107$, $df_1 = k = 3$, $df_2 = n - (k + 1) = 107 - 4 = 103$, $R^2 = 0.686$

$$F = \frac{R^2/k}{(1-R^2)/[n-(k+1)]} = \frac{.686/3}{(1-.686)/103} = 75.01$$

From Appendix Table VII, 0.001 > P-value.

Since the P-value is less than α, the null hypothesis is rejected. The data suggests that there is a useful linear relationship between exam score and at least one of the three predictor variables.

c   The 95% confidence interval for β_2 is

$3.369 \pm (1.98)(0.456) \Rightarrow 3.369 \pm 0.903 \Rightarrow (2.466, 4.272)$.

d   The point prediction would be $2.178 + 0.469(75) + 3.369(8) + 3.054(2.8) = 72.856$.

e   The prediction interval would be $72.856 \pm (1.98) \sqrt{s_e^2 + (1.2)^2}$.

To determine s_e^2, proceed as follows. From the definition of R^2, it follows that SSResid $= (1-R^2)$SSTo. So SSResid $= (1-0.686)(10200) = 3202.8$.

Then, $s_e^2 = \dfrac{3202.8}{103} = 31.095$.

The prediction interval becomes

$72.856 \pm (1.98) \sqrt{31.095 + (1.2)^2} \Rightarrow 72.856 \pm (1.98)(5.704)$
$\Rightarrow 72.856 \pm 11.294 \Rightarrow (61.562, 84.150)$.

14.43 $H_0: \beta_3 = 0 \quad H_a: \beta_3 \neq 0$

$\alpha = 0.05$

$t = \dfrac{b_3}{s_{b_3}}$ with d.f. $= 363$

$t = \dfrac{.00002}{.000009} = 2.22$

P-value $= 2$(area under the 363 df t curve to the right of 2.22) $\approx 2(0.014) = 0.028$.

Since the P-value is less than α, the null hypothesis is rejected. The conclusion is that the inclusion of the interaction term is important.

14.45   **a** $H_0: \beta_1 = \beta_2 = \beta_3 = 0$
H_a: At least one of the three β_i's is not zero.

$\alpha = 0.05$

Test statistic: $F = \dfrac{\text{SSRegr}/k}{\text{SSResid}/[n-(k+1)]}$

$$F = \frac{5073.4/3}{1854.1/6} = 5.47$$

From Appendix Table VII, $0.05 > $ P-value > 0.01.

Since the P-value is less than α, the null hypothesis is rejected. The data suggests that the model has utility for predicting discharge amount.

b H_o: $\beta_3 = 0$ H_a: $\beta_3 \neq 0$

$\alpha = 0.05$

The test statistic is: $t = \dfrac{b_3}{s_{b_3}}$ with df. $= 6$.

$$t = \frac{8.4}{199} = 0.04$$

P-value $= 2$(area under the 6 df t curve to the right of 0.04) $\approx 2(0.48) = 0.96$.

Since the P-value exceeds α, the null hypothesis is not rejected. The data suggests that the interaction term is not needed in the model, if the other two independent variables are in the model.

c   No. The model utility test is testing all variables simultaneously (that is, as a group). The t test is testing the contribution of an individual predictor when used in the presence of the remaining predictors. Results indicate that, given two out of the three predictors are included in the model, the third predictor may not be necessary.

14.47   The point prediction for mean phosphate adsorption when $x_1 = 160$ and $x_2 = 39$ is at the midpoint of the given interval. So the value of the point prediction is $(21.40 + 27.20)/2 = 24.3$. The t critical value for a 95% confidence interval is 2.23. The standard error for the point prediction is equal to $(27.20 - 21.40)/2(2.23) = 1.30$. The t critical value for a 99% confidence interval is 3.17. Therefore, the 99% confidence interval would be

$$24.3 \pm (3.17)(1.3) \Rightarrow 24.3 \pm 4.121 \Rightarrow (20.179, 28.421).$$

14.49   **a** H_o: $\beta_1 = \beta_2 = 0$
          H_a: At least one of the two β_i's is not zero.

$\alpha = 0.05$

Test statistic: $F = \dfrac{\text{SSRegr} / k}{\text{SSResid} / [n - (k+1)]}$

$$F = \frac{237.52/2}{26.98/7} = 30.81$$

From Appendix Table VII, 0.001 > P-value.

Since the P-value is less than α, the null hypothesis is rejected. The data suggests that the fitted model is useful for predicting plant height.

b $\alpha = 0.05$. From the MINITAB output the t-ratio for b_1 is 6.57, and the t-ratio for b_2 is -7.69. The P-values for the testing $\beta_1 = 0$ and $\beta_2 = 0$ would be twice the area under the 7 df t curve to the right of 6.57 and 7.69, respectively. From Appendix Table IV, the P-values are found to be practically zero. Both hypotheses would be rejected. The data suggests that both the linear and quadratic terms are important.

c   The point estimate of the mean y value when $x = 2$ is
$\hat{y} = 41.74 + 6.581(2) - 2.36(4) = 45.46$.

The 95% confidence interval is $45.46 \pm (2.37)(1.037) \Rightarrow 45.46 \pm 2.46 \Rightarrow (43.0, 47.92)$.

With 95% confidence, the mean height of wheat plants treated with $x = 2$ ($10^2 = 100$ uM of Mn) is estimated to be between 43 and 47.92 cm.

d   The point estimate of the mean y value when $x = 1$ is
$\hat{y} = 41.74 + 6.58(1) - 2.36(1) = 45.96$.

The 90% confidence interval is $45.96 \pm (1.9)(1.031) \Rightarrow 45.96 \pm 1.96 \Rightarrow (44.0, 47.92)$.

With 90% confidence, the mean height of wheat plants treated with $x = 1$ ($10 = 10$ uM of Mn) is estimated to be between 44 and 47.92 cm.

Exercises 14.51 – 14.62

14.51   One possible way would have been to start with the set of predictor variables consisting of all five variables, along with all quadratic terms, and all interaction terms. Then, use a selection procedure like backward elimination to arrive at the given estimated regression equation.

14.53 The model using the three variables x_3, x_9, x_{10} appears to be a good choice. It has an adjusted R^2 which is only slightly smaller than the largest adjusted R^2. This model is almost as good as the model with the largest adjusted R^2 but has two less predictors.

14.55 **a** The model has 9 predictors.

H_o: $\beta_1 = \beta_2 = \beta_3 = \beta_4 = \beta_5 = \beta_6 = \beta_7 = \beta_8 = \beta_9 = 0$
H_a: at least one among β_1, β_2, β_3, β_4, β_5, β_6, β_7, β_8, β_9 is not zero

$\alpha = 0.05$ (for illustration).

$$F = \frac{R^2/k}{(1-R^2)/[n-(k+1)]}$$

$n = 1856$, $df_1 = k = 9$, $df_2 = n - (k + 1) = 1856 - 10 = 1846$.

$R^2 = = 0.3092$.

$$F = \frac{R^2/k}{(1-R^2)/[n-(k+1)]} = \frac{0.3092/9}{(1-0.3092)/1846} = 91.8$$

From Appendix Table VII, $0.001 > $ P-value.

Since the P-value is less than α, the null hypothesis is rejected. There does appear to be a useful linear relationship between ln(blood cadmium level) and at least one of the nine predictors.

b If a backward elimination procedure was followed in the stepwise regression analysis, then the statements in the paper suggest that all variables except daily cigarette consumption and alcohol consumption were eliminated from the model. Of the two predictors left in the model, cigarette consumption would have a larger t-ratio than alcohol consumption.

There is an alternative procedure called the forward selection procedure which is available in most statistical software packages including MINITAB. According to this method one starts with a model having only the intercept term and enters one predictor at a time into the model. The predictor explaining most of the variance is entered first. The second predictor entered into the model is the one that explains most of the remaining variance, and so on. If the forward selection method was followed in the current problem then the statements in the paper would suggest that the variable to enter the model first is daily cigarette consumption and the next variable to enter the model is alcohol consumption. No further predictors were entered into the model.

14.57   **a**   Yes, they do show a similar pattern.

b   Standard error for the estimated coefficient of log of sales = (estimated coefficient)/t-ratio = 0.372/6.56 = 0.0567.

c   The predictor with the smallest (in magnitude) associated t-ratio is Return on Equity. Therefore it is the first candidate for elimination from the model. It has a t-ratio equal to 0.33 which is much less than $t_{out} = 2.0$. Therefore the predictor Return on Equity would be eliminated from the model if a backward elimination method is used with $t_{out} = 2.0$.

d   No. For the 1992 regression, the first candidate for elimination when using a backward elimination procedure is CEO Tenure since it has the smallest t-ratio (in magnitude).

e   We test H_o: Coefficient of Stock Ownership is equal to 0 versus H_a: Coefficient of Stock Ownership is less than 0. The t-ratio for this test is -0.58. Using Table IV from the appendix we find that the area to the left of the 153 d.f. t curve is approximately 0.3. So, the P-value for the test is approximately 0.3. Using MINITAB we find that the exact P-value is 0.2814.

14.59   Using MINITAB, the best model with k variables has been found and summary statistics from each are given below.

Number of Variables	Variables Included	R^2	Adjusted R^2	Cp
1	x_4	0.824	0.819	14.0
2	x_2, x_4	0.872	0.865	2.9
3	x_2, x_3, x_4	0.879	0.868	3.1
4	x_1, x_2, x_3, x_4	0.879	0.865	5.0

The best model, using the procedure of minimizing Cp, would use variables x_2, x_4. Hence, the set of predictor variables selected here is not the same as in problem **14.58**.

14.61   Using MINITAB, the best model with k variables has been found and summary statistics for each are given below.

k	Variables Included	R^2	Adjusted R^2	Cp
1	x_4	0.067	0.026	5.8
2	x_2, x_4	0.111	0.031	6.6
3	x_1, x_3, x_4	0.221	0.110	5.4
4	x_1, x_3, x_4, x_5	0.293	0.151	5.4
5	x_1, x_2, x_3, x_4, x_5	0.340	0.166	6.0

It appears that the model using x_1, x_3, x_4 is the best model, using the criterion of minimizing Cp.

Exercises 14.636 – 14.73

14.63   **a** $H_o: \beta_1 = \beta_2 = \beta_3 = \ldots = \beta_{11} = 0$
H_a: at least one among β_i's is not zero

$\alpha = 0.01$

$$F = \frac{R^2/k}{(1-R^2)/[n-(k+1)]}$$

$n = 88, \ df_1 = k = 11, \ df_2 = n - (k + 1) = 88 - 12 = 76$

$$F = \frac{R^2/k}{(1-R^2)/[n-(k+1)]} = \frac{.64/11}{(1-.64)/76} = \frac{.058182}{.004737} = 12.28$$

Appendix Table VII does not have entries for $df_1 = 11$, but using $df_1 = 10$ it can be determined that $0.001 > $ P-value.

Since the P-value is less than α, the null hypothesis is rejected. There does appear to be a useful linear relationship between y and at least one of the predictors.

b $\text{Adjusted R}^2 = 1 - \left[\dfrac{n-1}{n-(k+1)}\right]\dfrac{\text{SSResid}}{\text{SSTo}}$

To calculate adjusted R^2, we need the values for SSResid and SSTo. From the information given, we obtain:

$$s_e^2 = (5.57)^2 \Rightarrow 31.0249 = \frac{\text{SSResid}}{88-12} \Rightarrow \text{SSResid} = 76(31.0249) = 2357.8924$$

$$R^2 = .64 \Rightarrow .64 = 1 - \frac{\text{SSResid}}{\text{SSTo}} \Rightarrow .64 = 1 - \frac{2357.8924}{SSTol} \Rightarrow \frac{2357.8924}{SSTo} = .36$$

$$\Rightarrow SSTo = \frac{2357.8924}{.36} = 6549.7011.$$

So, $Adjusted\ R^2 = 1 - \left[\frac{n-1}{n-(k+1)}\right]\frac{\text{SSResid}}{\text{SSTo}} = 1 - \frac{87}{76}\left(\frac{2357.8924}{6549.7011}\right) = 1 - .4121 = .5879.$

c $\text{t-ratio} = \frac{b_1 - 0}{s_{b_1}} = 3.08 \Rightarrow s_{b_1} = \frac{b_1}{3.08} = \frac{.458}{3.08} = .1487$

$b_1 \pm (t\ \text{critical})s_{b_1} \Rightarrow .458 \pm (2.00)(.1487) \Rightarrow .458 \pm .2974 \Rightarrow (.1606, .7554)$

From this interval, we estimate the value of β_1 to be between -0.1606 and 0.7554.

d   Many of the variables have t-ratios that are close to zero. The one with the smallest t in absolute value is x_9: certificated staff-pupil ratio. For this reason, I would eliminate x_9 first.

e $H_o: \beta_3 = \beta_4 = \beta_5 = \beta_6 = 0$
        $H_a:$ at least one among $\beta_3, \beta_4, \beta_5, \beta_6$ is non-zero.

None of the procedures presented in this chapter could be used. The two procedures presented tested "all variables as a group" or "a single variable's contribution".

14.65 **a**

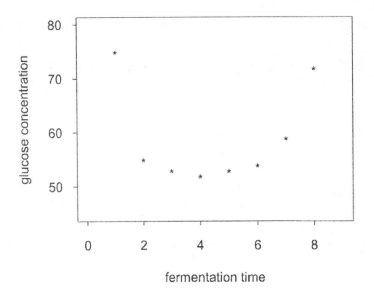

fermentation time

Based on this scatterplot a quadratic model in x is suggested.

b $H_o: \beta_1 = \beta_2 = 0$
$H_a:$ At least one of the two β_i's is not zero.

$\alpha = 0.05$

Test statistic: $F = \dfrac{SSRegr \, / \, k}{SSResid \, / \, [n - (k+1)]}$

$F = \dfrac{525.11 / 2}{61.77 / 5} = 21.25$

From Appendix Table VII, $0.01 > P\text{-value} > 0.001$.

Since the P-value is less than α, the null hypothesis is rejected. The data suggests that the quadratic model is useful for predicting glucose concentration.

c $H_o: \beta_2 = 0$ $H_a: \beta_2 \neq 0$

$\alpha = 0.05$

The test statistic is: $t = \dfrac{b_2}{s_{b_2}}$ with df. $= 5$.

$$t = \frac{1.7679}{.2712} = 6.52$$

P-value = 2(area under the 5 df t curve to the right of 6.52) ≈ 2(0) = 0.

Since the P-value is less than α, the null hypothesis is rejected. The data suggests that the quadratic term cannot be eliminated from the model.

14.67 When n = 21 and k = 10, Adjusted $R^2 = 1 - 2$(SSResid/SSTo).
Then Adjusted $R^2 < 0 \Rightarrow \frac{1}{2} <$ SSResid / SSTo = $1 - R^2 \Rightarrow 1/2 > R^2$.
Hence, when n = 21 and k = 10, Adjusted R^2 will be negative for values of R^2 less than 0.5.

14.69 First, the model using all four variables was fit. The variable age at loading (x_3) was deleted because it had the t-ratio closest to zero and it was between –2 and 2. Then, the model using the three variables x_1, x_2, and x_4 was fit. The variable time (x_4) was deleted because its t-ratio was closest to zero and was between –2 and 2. Finally, the model using the two variables x_1 and x_2 was fit. Neither of these variables could be eliminated since their t-ratios were greater than 2 in absolute magnitude. The final model then, includes slab thickness (x_1) and load (x_2). The predicted tensile strength for a slab that is 25 cm thick, 150 days old, and is subjected to a load of 200 kg for 50 days is $\hat{y} = 13 - 0.487(25) + 0.0116(200) = 3.145$.

14.71 **a**

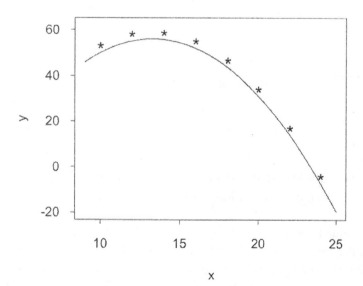

b The claim is very reasonable because 14 is close to where the smooth curve has its highest value.

14.73 a   Output from MINITAB is given below.

The regression equation is: $Y = 1.56 + .0237 X1 - 0.000249 X2$.

Predictor	Coef	Stdev	t-ratio	p
Constant	1.56450	0.07940	19.70	0.000
X1	0.23720	0.05556	4.27	0.000
X2	−0.00024908	0.00003205	−7.77	0.000

$s = 0.05330$ R-sq = 86.5%   R-sq(adj) = 85.3%

Analysis of Variance

SOURCE	DF	SS	MS	F	p
Regression	2	0.40151	0.20076	70.66	0.000
Error	22	0.06250	0.00284		
Total	24	0.46402			

b H_o: $\beta_1 = \beta_2 = 0$
H_a: At least one of the two β_i's is not zero.

$\alpha = 0.05$

Test statistic: $F = \dfrac{\text{SSRegr} / k}{\text{SSResid} / [n - (k+1)]}$

$F = \dfrac{.40151/2}{.0625/22} = 70.67$

From the Minitab output, the P-value associated with the F test is practically zero. Since the P-value is less than α, the null hypothesis is rejected.

c   The value for R^2 is 0.865. This means that 86.5% of the total variation in the observed values for profit margin has been explained by the fitted regression equation. The value for s_e is 0.0533. This means that the typical deviation of an observed value from the predicted value is 0.0533, when predicting profit margin using this fitted regression equation.

d   No. Both variables have associated t-ratios that exceed 2 in absolute magnitude. Hence, neither can be eliminated from the model.

e

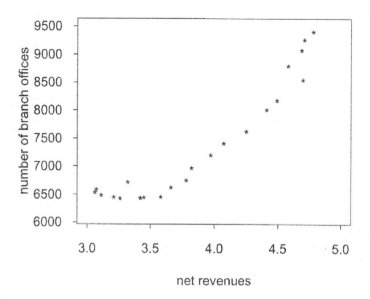

net revenues

There do not appear to be any influential observations. However, there is substantial evidence of multicollinearity. The plot shows a pronounced linear relationship between x_1 and x_2. This is evidence of multicollinearity between x_1 and x_2.

Chapter 15

Exercises 15.1 – 15.18

15.1   **a** $0.001 < \text{P-value} < 0.01$

        b $\text{P-value} > 0.10$

        c $\text{P-value} = 0.01$

        d $0.001 > \text{P-value}$

        e $0.05 < \text{P-value} < 0.10$

        f $0.01 < \text{P-value} < 0.05$ (Using $df_2 = 40$ and $df_2 = 60$ tables).

15.3   **a**   Let μ_1, μ_2, μ_3 and μ_4 denote the true average length of stay in a hospital for health plans 1, 2, 3 and 4 respectively.

           H_0: $\mu_1 = \mu_2 = \mu_3 = \mu_4$
           H_a: At least two of the four μ_i's are different.

        b $df_1 = 4 - 1 = 3$ $df_2 = 32 - 4 = 28$ $\alpha = 0.01$

           From Appendix Table VII, $0.05 > \text{P-value} > 0.01$, Since the P-value exceeds α, H_0 is not rejected. Hence, it would be concluded that the average length of stay in the hospital is
the same for the four health plans.

        c $df_1 = 4 - 1 = 3$ $df_2 = 32 - 4 = 28$ $\alpha = 0.01$

           From Appendix Table VII, $0.05 > \text{P-value} > 0.01$. Since the P-value exceeds α, H_0 is not rejected. Therefore, the conclusion would be the same.

15.5   **a**   The required boxplot obtained using MINITAB is shown below. Price per acre values appear to be similar for 1996 and 1997 but 1998 values are higher. The mean price per acre values for each year are also plotted as a solid square within each box plot.

b   Let μ_i denote the mean price per acre for vineyards in year i (i = 1, 2, 3).
H_0: $\mu_1 = \mu_2 = \mu_3$
H_a: At least two of the three μ_i's are different.

$\alpha = 0.01$

Test statistic: $F = \dfrac{MSTr}{MSE}$

$df_1 = k - 1 = 2$ $df_2 = N - k = 15\text{-}3 = 12.$

$\bar{x}_1 = 35600, \bar{x}_2 = 36000, \bar{x}_3 = 43600$

$$\bar{\bar{x}} = \frac{[5(35600) + 5(36000) + 5(43600)]}{(15)} = 38400$$

MSTr = [5(35600 − 38400)² + 5(36000 − 38400)² + 5(43600 − 38400)²]/2 = 101600000

$s_1 = 3847.077, s_2 = 3807.887, s_3 = 3911.521$
MSE = [(5-1)(3847.077)² + (5-1)(3807.887)² + (5-1)(3911.521)²]/12 = 14866667

310

$$F = \frac{\text{MSTr}}{\text{MSE}} = \frac{101600000}{14866667} = 6.83$$

From Appendix Table VII, $0.05 > \text{P-value} > 0.01$.

Since the P-value exceeds α, the null hypothesis is not rejected. At a significance level of $\alpha = 0.01$, the data does not support the claim that the true mean price per acre for the three years under consideration are different.

15.7 Let μ_i denote the mean level of chlorophyll concentration for plants in variety i (i = 1, 2, 3, 4).
$H_o: \mu_1 = \mu_2 = \mu_3 = \mu_4$
H_a: At least two of the four μ_i's are different.

$\alpha = 0.05$

Test statistic: $F = \dfrac{\text{MSTr}}{\text{MSE}}$

$df_1 = k - 1 = 3 \quad df_2 = N - k = 16.$

$$\overline{\overline{x}} = \frac{[5(.3) + 5(.24) + 4(.41) + 6(.33)]}{(20)} = 0.316$$

$\text{MSTr} = [5(0.3 - 0.316)^2 + 5(0.24 - 0.316)^2 + 4(0.41 - 0.316)^2 + 6(0.33 - 0.316)^2]/3 = 0.06668/3 = 0.022227$

$$F = \frac{\text{MSTr}}{\text{MSE}} = \frac{.022227}{.013} = 1.71$$

From Appendix Table VII, P-value > 0.10.

Since the P-value exceeds α, the null hypothesis is not rejected. The data does not suggest that true mean chlorophyll concentration differs for the four varieties.

15.9 μ_1 : true average importance rating of speed for owners of American cars.
μ_2 : true average importance rating of speed for owners of German cars.
μ_3 : true average importance rating of speed for owners of Japanese cars.

$H_o: \mu_1 = \mu_2 = \mu_3$
H_a: at least two among μ_1, μ_2, μ_3 are not equal.

$\alpha = 0.05$

$$F = \frac{MSTr}{MSE}$$

$df_1 = 2 \quad df_2 = (58 + 38 + 59) - 3 = 155 - 3 = 152$

$T = 58(3.05) + 38(2.87) + 59(2.67) = 443.49$

$$\bar{\bar{x}} = \frac{T}{N} = \frac{443.49}{155} = 2.8612$$

$SSTr = 58(3.05 - 2.8612)^2 + 38(2.87 - 2.8612)^2 + 59(2.67 - 2.8612)^2$
$\quad = 2.066870 + 0.002925 + 2.157471 = 4.227267$

$$MSTr = \frac{4.227267}{2} = 2.113634$$

$$MSE = \frac{459.04}{152} = 3.02$$

$$F = \frac{MSTr}{MSE} = \frac{2.113634}{3.02} = .70$$

From Appendix Table VII, P-value > 0.10.

Since the P-value exceeds α, H_o is not rejected. It is plausible that the true average importance of speed rating is the same for the three groups.

15.11   Let μ_i denote the mean dry weight for concentration level i (i = 1, 2, ..., 10).
$H_o: \mu_1 = \mu_2 = ... = \mu_{10}$
$H_a:$ At least two of the ten μ_i's are different.

$\alpha = 0.05$

Test statistic: $F = \frac{MSTr}{MSE}$

$df_1 = k - 1 = 9 \quad df_2 = N - k = 30.$

$$F = \frac{MSTr}{MSE} = 1.895$$

From Appendix Table VII, 0.10 > P-value > 0.05.

Since the P-value exceeds α, the null hypothesis is not rejected. The data are consistent with the hypothesis that the true mean dry weight does not depend on the level of concentration.

15.13

Source of Variation	Degrees of Freedom	Sum of Squares	Mean Square	F
Treatments	3	75081.72	25027.24	1.70
Error	16	235419.04	14713.69	
Total	19	310500.76		

Let μ_i denote the mean number of miles to failure for brand i sparkplugs (i = 1, 2, 3, 4).
H_0: $\mu_1 = \mu_2 = \mu_3 = \mu_4$
H_a: At least two of the four μ_i's are different.

$\alpha = 0.05$

Test statistic: $F = \dfrac{MSTr}{MSE}$

$df_1 = k - 1 = 3$ $df_2 = N - k = 16$.   From the ANOVA table, F = 1.70.

From Appendix Table VII, P-value > 0.10.

Since the P-value exceeds α, the null hypothesis is not rejected. The data are consistent with the hypothesis that there is no difference between the mean number of miles to failure for the four brands of sparkplugs.

15.15 Computations: $\bar{\bar{x}} = [96(2.15) + 34(2.21) + 86(1.47) + 206(1.69)]/422$
$= 756.1/422 = 1.792$

$MSTr = [96(2.15 - 1.792)^2 + 34(2.21 - 1.792)^2 + 86(1.47 - 1.792)^2$
$+ 206(1.69 - 1.792)^2]/3 = 29.304/3 = 9.768$

$MSE = \dfrac{MSTr}{F} = \dfrac{9.768}{2.56} = 3.816$

Source of Variation	Degrees of Freedom	Sum of Squares	Mean Square	F
Treatments	3	29.304	9.768	2.56
Error	418	1595.088	3.816	
Total	421	1624.392		

Let μ_i denote the mean number of hours per month absent for employees of group i
(i = 1, 2, 3, 4).

$H_0: \mu_1 = \mu_2 = \mu_3 = \mu_4$
H_a: At least two of the four μ_i's are different.

$\alpha = 0.01$

Test statistic: $F = \dfrac{MSTr}{MSE}$

$df_1 = k - 1 = 3$ $df_2 = N - k = 418$. From the ANOVA table, F = 2.56.

From Appendix Table VII, $0.10 > P\text{-value} > 0.05$.

Since the P-value exceeds α, the null hypothesis is not rejected. The data are consistent with the hypothesis that there is no difference between the mean number of hours per month absent for employees in the four groups.

15.17 Let μ_1, μ_2, and μ_3 denote the true mean fog indices for *Scientific America*, *Fortune*, and *New Yorker*, respectively.
$H_0: \mu_1 = \mu_2 = \mu_3$
H_a: At least two of the three μ_i's are different.

$\alpha = 0.01$

Test statistic: $F = \dfrac{MSTr}{MSE}$
$df_1 = k - 1 = 2$ $df_2 = N - k = 15$.

Computations: $\bar{\bar{x}} = 9.666$

Magazine	Mean	Standard Deviation
S.A.	10.968	2.647
F	10.68	1.202
N.Y.	7.35	1.412

$MSTr = [6(10.968 - 9.666)^2 + 6(10.68 - 9.666)^2 + 6(7.35 - 9.666)^2]/2 = 48.524/2 = 24.262$

$MSE = [(2.647)^2 + (1.202)^2 + (1.412)^2]/3 = 10.443/3 = 3.4812$

Source of Variation	Degrees of Freedom	Sum of Squares	Mean Square	F
Treatments	2	48.524	24.2620	6.97
Error	15	55.218	3.4812	
Total	17	100.742		

$$F = \frac{MSTr}{MSE} = \frac{24.262}{3.4812} = 6.97$$

From Appendix Table VII, $0.01 > P\text{-value} > 0.001$.

Since the P-value is less than α, the null hypothesis is rejected. The data suggests that there is a difference between at least two of the mean fog index levels for advertisements appearing in the three magazines.

Exercises 15.19 – 15.29

15.19 Since the intervals for $\mu_1 - \mu_2$ and $\mu_1 - \mu_3$ do not contain zero, μ_1 and μ_2 are judged to be different and μ_1 and μ_3 are judged to be different. Since the interval for $\mu_2 - \mu_3$ contains zero, μ_2 and μ_3 are judged not to be different. Hence, statement (iii) best describes the relationship between μ_1, μ_2, and μ_3.

15.21 μ_1 differs from μ_2; μ_1 differs from μ_3; μ_1 differs from μ_5; μ_2 differs from μ_4; μ_3 differs from μ_4; μ_4 differs from μ_5.

From the data given the following means were computed:

$$\bar{x}_1 = 16.35 \quad \bar{x}_2 = 11.63 \quad \bar{x}_3 = 10.5 \quad \bar{x}_4 = 14.96 \quad \bar{x}_5 = 12.3$$

Fabric	3	2	5	4	1
$\bar{x}$	10.5	11.63	12.3	14.96	16.35

15.23 $k = 4$ Error df $= (5 + 5 + 4 + 6) - 4 = 16$

From Appendix Table VIII, $q = 4.05$ for 95% confidence.

$$\mu_1 - \mu_2 : (.30 - .24) \pm 4.05 \sqrt{\frac{.013}{2}\left(\frac{1}{5} + \frac{1}{5}\right)} \Rightarrow .06 \pm .2065 \Rightarrow (-.1465, .2665)$$

$$\mu_1 - \mu_3 : (.30 - .41) \pm 4.05 \sqrt{\frac{.013}{2}\left(\frac{1}{5} + \frac{1}{4}\right)} \Rightarrow -.11 \pm .2190 \Rightarrow (-.3190, .1090)$$

$$\mu_1 - \mu_4 : (.30 - .33) \pm 4.05 \sqrt{\frac{.013}{2}\left(\frac{1}{5} + \frac{1}{6}\right)} \Rightarrow -.03 \pm .1977 \Rightarrow (-.2277, .1677)$$

$$\mu_2 - \mu_3 : (.24 - .41) \pm 4.05 \sqrt{\frac{.013}{2}\left(\frac{1}{5} + \frac{1}{4}\right)} \Rightarrow -.17 \pm .2190 \Rightarrow (-.3890, .0490)$$

$$\mu_2 - \mu_4 : (.24 - .33) \pm 4.05 \sqrt{\frac{.013}{2}\left(\frac{1}{5} + \frac{1}{6}\right)} \Rightarrow -.09 \pm .1977 \Rightarrow (-.2877, .1077)$$

$$\mu_3 - \mu_4 : (.41 - .33) \pm 4.05 \sqrt{\frac{.013}{2}\left(\frac{1}{4} + \frac{1}{6}\right)} \Rightarrow .08 \pm .2108 \Rightarrow (-.1308, .2908)$$

There are no pairwise differences.

Variety	2. RO	1. BI	4. TO	3. WA
Mean	0.24	0.30	0.33	0.41

No differences are detected. This conclusion is in agreement with the results of the F test done in problem 15.7.

$\alpha = 0.05$ (for illustration)

Test statistic: $F = \dfrac{MSTr}{MSE}$

$df_1 = k - 1 = 3$ $df_2 = N - k = 76$. From the ANOVA table, $F = 45.6432$.

From Appendix Table VII, $0.001 >$ P-value.

Since the P-value is smaller than α, the null hypothesis is rejected. The data suggests that the mean number of germinating seeds for the four treatments are not all the same.

b   Let μ_2 = the mean number of germinating seeds corresponding to lizard dung and μ_3 = the mean number of germinating seeds corresponding to bird dung. We wish to test $H_o : \mu_2 = \mu_3$ versus $H_a: \mu_2 > \mu_3$. Since only two means are being compared we do not have to carry out a multiple comparisons procedure.

Test statistic: $t = \dfrac{(\bar{x}_2 - \bar{x}_3) - 0}{s.e.(\bar{x}_2 - \bar{x}_3)}$

$\bar{x}_2 - \bar{x}_3 = 2.35 - 1.70 = 0.65$.

s.e. of $\bar{x}_2 - \bar{x}_3 = \sqrt{MSE\left(\dfrac{1}{20} + \dfrac{1}{20}\right)} = \sqrt{(0.098225)\left(\dfrac{1}{20} + \dfrac{1}{20}\right)} = 0.0991$.

t-statistic = $\dfrac{0.65}{0.0991} = 6.56$.

The area to the right of 6.56 under the 76 d.f. t curve is ≈ 0. Hence the null hypothesis is rejected and we conclude that the mean number of germinating seeds for lizard dung is greater than that for bird dung.

15.25

Group	Simultaneous	Sequential	Control
Mean	$\bar{x}_3$	$\bar{x}_2$	$\bar{x}_1$

15.27   The mean water loss when exposed to 4 hours fumigation is different from all other means. The mean water loss when exposed to 2 hours fumigation is different from that for levels 16 and 0, but not 8. The mean water losses for duration 16, 0, and 8 hours are not different from one another. No other differences are significant.

15.29   **a** $H_o: \mu_1 = \mu_2 = \mu_3 = \mu_4 = \mu_5$
H_a: At least two of the five μ_i's are different.

$\alpha = 0.05$

Test statistic: $F = \dfrac{MSTr}{MSE}$

$df_1 = k - 1 = 4$ $df_2 = N - k = 15$.

From the data, the following statistics were calculated.

Hormone	1	2	3	4	5
n	4	4	4	4	4
mean	12.75	17.75	17.50	11.50	10.00
Total	51	71	70	46	40
variance	17.583	12.917	8.333	21.667	15.333

$$\bar{\bar{x}} = \frac{T}{N} = \frac{278}{20} = 13.9$$

$SSTr = 4(12.75 - 13.9)^2 + 4(17.75 - 13.9)^2 + 4(17.50 - 13.9)^2 + 4(11.5 - 13.9)^2$
$\qquad + 4(10 - 13.9)^2$
$\qquad = 5.29 + 59.29 + 51.84 + 23.04 + 60.840 = 200.3$

$SSE = 3(17.583) + 3(12.917) + 3(8.333) + 3(21.667) + 3(15.333) = 227.499$

$df_1 = k - 1 = 5 - 1 = 4$ $df_2 = (4 + 4 + 4 + 4 + 4) - 5 = 15$

$$MSTr = \frac{SSTr}{k-1} = \frac{200.3}{4} = 50.075$$

$$MSE = \frac{SSE}{N-k} = \frac{227.499}{15} = 15.1666$$

$$F = \frac{MSTr}{MSE} = \frac{50.075}{15.1666} = 3.30$$

From Appendix Table VII, $0.05 > $ P-value > 0.01.

Since the P-value is less than α, H_o is rejected. The data supports the conclusion that the mean plant growth is not the same for all five growth hormones.

b $k = 5$ Error df $= 15$

From Appendix Table VIII, q = 4.37.

Since the sample sizes are the same, the ± factor is the same for each comparison.

$$4.37\sqrt{\frac{15.1666}{2}\left(\frac{1}{4}+\frac{1}{4}\right)} = 8.51$$

$\mu_1 - \mu_2$: $(12.75 - 17.75) \pm 8.51 \Rightarrow -5 \pm 8.51 \Rightarrow (-13.51, 3.51)$

$\mu_1 - \mu_3$: $(12.75 - 17.50) \pm 8.51 \Rightarrow -4.75 \pm 8.51 \Rightarrow (-13.26, 3.76)$

$\mu_1 - \mu_4$: $(12.75 - 11.5) \pm 8.51 \Rightarrow 1.25 \pm 8.51 \Rightarrow (-7.26, 9.76)$

$\mu_1 - \mu_5$: $(12.75 - 10) \pm 8.51 \Rightarrow 2.75 \pm 8.51 \Rightarrow (-5.76, 11.26)$

$\mu_2 - \mu_3$: $(17.75 - 17.50) \pm 8.51 \Rightarrow 0.25 \pm 8.51 \Rightarrow (-8.26, 8.76)$

$\mu_2 - \mu_4$: $(17.75 - 11.5) \pm 8.51 \Rightarrow 6.25 \pm 8.51 \Rightarrow (-2.26, 14.76)$

$\mu_2 - \mu_5$: $(17.75 - 10) \pm 8.51 \Rightarrow 7.75 \pm 8.51 \Rightarrow (-0.76, 16.26)$

$\mu_3 - \mu_4$: $(17.5 - 11.5) \pm 8.51 \Rightarrow 6.0 \pm 8.51 \Rightarrow (-2.51, 14.51)$

$\mu_3 - \mu_5$: $(17.5 - 10) \pm 8.51 \Rightarrow 7.5 \pm 8.51 \Rightarrow (-1.01, 16.01)$

$\mu_4 - \mu_5$: $(11.5 - 10) \pm 8.51 \Rightarrow 1.5 \pm 8.51 \Rightarrow (-7.01, 10.01)$

No significant differences are determined using the T-K method.

Exercises 15.30 – 15.38

15.31   a

Source of Variation	Degrees of Freedom	Sum of Squares	Mean Square	F
Treatments	2	11.7	5.85	0.37
Blocks	4	113.5	28.375	
Error	8	125.6	15.7	
Total	14	250.8		

b H_0: The mean appraised value does not depend on which assessor is doing the appraisal.
H_a: The mean appraised value does depend on which assessor is doing the appraisal.

$\alpha = 0.05$

Test statistic: $F = \dfrac{MSTr}{MSE}$

$df_1 = k - 1 = 2$ $df_2 = (k-1)(l-1) = 8$. From the ANOVA table, $F = 0.37$.

From Appendix Table VII, P-value > 0.10.

Since the P-value exceeds α, the null hypothesis is not rejected. The mean appraised value does not seem to depend on which assessor is doing the appraisal.

15.33

Source of Variation	Degrees of Freedom	Sum of Squares	Mean Square	F
Treatments	2	3.97	1.985	79.4
Blocks	7	0.2503	0.0358	
Error	14	0.3497	0.025	
Total	23	4.57		

H_0: The mean energy use does not depend on the type of oven.
H_a: The mean energy use does depend on the type of oven.

$\alpha = 0.01$

Test statistic: $F = \dfrac{MSTr}{MSE}$

$df_1 = k - 1 = 2$ and $df_2 = (k-1)(l-1) = 14$. From the ANOVA table, $F = 79.4$.

From Appendix Table VII, $0.001 >$ P-value.

Since the P-value is less than α, the null hypothesis is rejected. The data suggests quite strongly that the mean energy use depends on the type of oven used.

15.35   **a**   Other environmental factors (amount of rainfall, number of days of cloudy weather, average daily temperature, etc.) vary from year to year. Using a randomized complete block helps control for variation in these other factors.

      b $SSTr = 3[(138.33 - 149)^2 + (152.33 - 149)^2 + (156.33 - 149)^2] = 536.00$

           $SSBl = 3[(173 - 149)^2 + (123.67 - 149)^2 + (150.33 - 149)^2] = 3658.13$

Source of Variation	Degrees of Freedom	Sum of Squares	Mean Square	F
Treatments	2	536.00	268.00	14.91
Blocks	2	3658.13	1829.07	
Error	4	71.87	17.97	
Total	8	4266.00		

H_0: The mean height does not depend on the rate of application of effluent.
H_a: The mean height does depend on the rate of application of effluent.

$\alpha = 0.05$

Test statistic: $F = \dfrac{MSTr}{MSE}$

$df_1 = k - 1 = 2$ and $df_2 = (k - 1)(l - 1) = 4$. From the ANOVA table, $F = 14.91$.

From Appendix Table VII, $0.05 > $ P-value > 0.01.

Since the P-value is less than α, the null hypothesis is rejected. The data suggests that the mean height of cotton plants does depend on the rate of application of effluent.

      c   Since the sample sizes are equal, the $\pm$ factor is the same for each comparison.

$k = 3$ Error df $= 4$ $q = 5.04$ (This value came from a more extensive table of critical values for the Studentized range distribution than the text provides.)

The $\pm$ factor is: $5.04\sqrt{\dfrac{17.97}{3}} = 12.33$

Application Rate	1	2	3
Mean	138.33	152.33	156.33

The mean height under an application rate of 350 differs from those using application rates of 440 and 515.

The mean height using application rates of 440 and 515 do not differ.

15.37 H_o: The mean height does not depend on the seed source.
H_a: The mean height does depend on the seed source.

$\alpha = 0.05$

Test statistic: $F = \dfrac{MSTr}{MSE}$

Source of Variation	Degrees of Freedom	Sum of Squares	Mean Square	F
Treatments	4	4.543	1.136	0.868
Blocks	3	7.862	2.621	
Error	12	15.701	1.308	
Total	19	28.106		

$df_1 = k - 1 = 4$, $df_2 = (k - 1)(l - 1) = 12$, and $F = 0.868$.

From Appendix Table VII, P-value > 0.10.
Since the P-value exceeds α, the null hypothesis is not rejected. The data are consistent with the hypothesis that mean height does not depend on the seed source.

Exercises 15.39 – 15.51

15.39   a

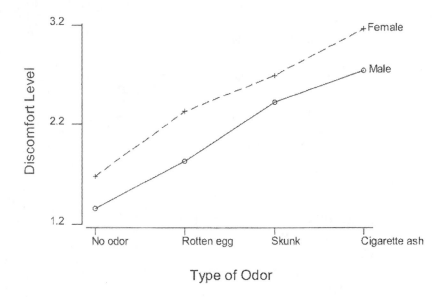

b   The graphs for males and females are very nearly parallel. There does not appear to be an interaction between gender and type of odor.

15.41   a   The plot does suggest an interaction between peer group and self-esteem. The change in average response, when changing from low to high peer group, is not the same for the low self-esteem group and the high self-esteem group. This is indicated by the non-parallel lines.

b   The change in the average response is greater for the low self-esteem group than it is for the high self-esteem group, when changing from low to high peer group interaction. Therefore, the authors are correct in their statement.

15.43

Source of Variation	Degrees of Freedom	Sum of Squares	Mean Square	F
Size (A)	2	0.088	0.044	4.00
Species (B)	1	0.048	0.048	4.363
Size by Species	2	0.048	0.024	2.18
Error	12	0.132	0.011	
Total	17	0.316		

H_o: There is no interaction between Size (A) and Species (B).
H_a: There is interaction between Size and Species.

$\alpha = 0.01$

The test statistic is: $F_{AB} = \dfrac{MSAB}{MSE}$.

$df_1 = 2$, $df_2 = 12$, and $F_{AB} = 2.18$

From Appendix Table VII, P-value > 0.10.

Since the P-value exceeds α, the null hypothesis is not rejected. The data are consistent with the hypothesis of no interaction between Size and Species. Hence, hypothesis tests on main effects will be done.

H_o: There are no size main effects.
H_a: There are size main effects.

$\alpha = 0.01$

The test statistic is: $F_A = \dfrac{MSA}{MSE}$.

$df_1 = 2$, $df_2 = 12$, and $F_A = 4.00$.

From Appendix Table VII, 0.05 > P-value > 0.01.

Since the P-value exceeds α, the null hypothesis of no size main effects is not rejected. When using $\alpha = 0.01$, the data support the conclusion that there are no differences between the mean preference indices for the three sizes of bass.

H_o: There are no species main effects.
H_a: There are species main effects.

$\alpha = 0.01$

The test statistic is: $F_B = \dfrac{MSB}{MSE}$.

$df_1 = 1$, $df_2 = 12.$, and $F_B = 4.363$

From Appendix Table VII, P-value > 0.05.

Since the P-value exceeds α, the null hypothesis of no species main effects is not rejected. At a significance level of $\alpha = 0.01$, the data are consistent with the hypothesis that there are no differences between the mean preference indices for the three species of bass.

15.45   **a** H_o: There is no interaction between Greek status (A) and Year (B).
H_a: There is interaction between Greek status and year in college.

$\alpha = 0.05$ (for illustration)

The test statistic is: $F_{AB} = \dfrac{MSAB}{MSE}$.

Factor A (Greek status) has 2 levels, so $df_A = 1$.

Factor B (Year in college) has 4 levels, so $df_B = 3$.

Hence $df_{AB} = (1)(3) = 3$.

Since the total number of observations is $75 + 57 = 132$, the d.f. for Total is 131 and the d.f. for Error is $131-(1+3+3) = 124$.

Hence, for the test of interaction we have

$df_1 = 3$, $df_2 = 124$, $F_{AB} = 0.70$.

From Appendix Table VII, P-value > 0.10.

Since the P-value is greater than α, the null hypothesis is not rejected. The data do not suggest existence of an interaction between Greek status and year in college. (Hence tests about main effects may be performed.)

b　　H_0: There is no Greek status main effect.
H_a: There is a Greek status main effect.

$\alpha = 0.05$ (for illustration)

The test statistic is: $F_A = \dfrac{MSA}{MSE}$.

$df_1 = 1$, $df_2 = 124$, and $F_A = 20.53$

From Appendix Table VII, $0.001 >$ P-value.

Since the P-value is smaller than α, the null hypothesis of no Greek status main effect is rejected. The data provide evidence to conclude that there is a difference between the mean "Self-esteem" scores of students belonging to a fraternity and those who do not belong to a fraternity.

c　　H_0: There are no year main effects.

H_a: There are year main effects.

$\alpha = 0.05$ (for illustration)

The test statistic is: $F_B = \dfrac{MSB}{MSE}$.

$df_1 = 3$, $df_2 = 124$, and $F_B = 2.59$

From Appendix Table VII, $0.10 >$ P-value > 0.05.

Since the P-value is greater than α, the null hypothesis of no year main effects is
　　　rejected.
The data do not suggest the existence of differences among the mean "Self-
　　　esteem"
scores of students belonging to different years in college.

15.47   The test for no interaction would have $df_1 = 2$, $df_2 = 120$, and $F_{AB} < 1$. From Appendix Table VII, P-value > 0.10. Since the P-value exceeds the α of 0.05, the null hypothesis of no interaction is not rejected. Since there appears to be no interaction, hypothesis tests on main effects are appropriate.

The test for no A main effects would have $df_1 = 2$, $df_2 = 120$, and $F_A = 4.99$. From Appendix Table VII, $0.05 > \text{P-value} > 0.01$. Since the P-value is less than α, the null hypothesis of no A main effects is rejected. The data suggests that the expectation of opportunity to cheat affects the mean test score.

The test for no B main effects would have $df_1 = 1$, $df_2 = 120$, and $F_B = 4.81$. From Appendix Table VII, $0.05 > \text{P-value} > 0.01$. Since the P-value is less than α, the null hypothesis of no B main effects is rejected. The data suggests that perceived payoff affects the mean test score.

15.49   **a**

Source of Variation	Degrees of Freedom	Sum of Squares	Mean Square	F
Race	1	857	857	5.57
Sex	1	291	291	1.89
Race by Sex	1	32	32	0.21
Error	36	5541	153.92	
Total	39	6721		

H_o: There is no interaction between race and sex.
H_a: There is interaction between race and sex.

$\alpha = 0.01$

Test statistic: $F_{AB} = \dfrac{MSAB}{MSE}$

$df_1 = 1$, $df_2 = 36$, and $F_{AB} = 0.21$. From Appendix Table VII, P-value > 0.10.

Since the P-value exceeds α, the null hypothesis of no interaction between race and sex is not rejected. Thus, hypothesis tests for main effects are appropriate.

b   H₀: There are no race main effects.
Hₐ: There are race main effects.

$\alpha = 0.01$

Test statistic: $F_A = \dfrac{MSA}{MSE}$

$df_1 = 1$, $df_2 = 36$, and $F_A = 5.57$. From Appendix Table VII, $0.05 > \text{P-value} > 0.01$

Since the P-value exceeds α, the null hypothesis of no race main effects is not rejected. The data are consistent with the hypothesis that the true average lengths of sacra do not differ for the two races.

c   H₀: There are no sex main effects.
Hₐ: There are sex main effects.

$\alpha = 0.01$

Test statistic: $F_B = \dfrac{MSB}{MSE}$

$df_1 = 1$, $df_2 = 36$, and $F_B = 1.89$. From Appendix Table VII, P-value > 0.10.

Since the P-value exceeds α, the null hypothesis of no sex main effects is not rejected. The data are consistent with the hypothesis that the true average lengths of sacra do not differ for males and females.

15.51 The following Anova table was obtained using the statistical software package Minitab.

Source of Variation	Degrees of Freedom	Sum of Squares	Mean Square	F
Rate	2	469.70	234.85	76.14
Soil Type	2	333.94	166.97	54.14
Error	4	12.34	3.08	
Total	8	815.98		

For each test, $df_1 = 2$, $df_2 = 4$. From Appendix Table VII, the P-value for the rate test is less than 0.001 and the P-value for the soil type is between 0.01 and 0.001. Since the P-values are less than $\alpha = 0.01$, both null hypotheses are rejected using a 0.01 level of significance. It appears that the total phosphorus uptake depends upon application rate as well as soil types.

$k = 3$, Error $df = 4$, $q = 5.04$ (This value came from a more extensive table of critical values for the Studentized range distribution than the text provides.)

The $\pm$ factor $= 5.04\sqrt{\dfrac{3.08}{3}} = 5.107$.

Soil Type	Ramiha	Konini	Wainui
Mean	10.217	20.957	24.557

The mean effect of soil types Komini and Wainui on total phosphorus uptake is the same, but differs for soil type Ramiha.

Exercises 15.52 – 15.63

15.53 H_o: $\mu_1 = \mu_2 = \mu_3 = \mu_4$
  H_a: at least two among μ_1, μ_2, μ_3, μ_4 are not equal.

$\alpha = 0.05$
$$F = \frac{MSTr}{MSE}$$

$N = 52$, $df_1 = 3$ $df_2 = 48$

$SSTr = SSTo - SSE = 682.10 - 506.19 = 175.91$

$$MSTr = \frac{175.91}{3} = 58.637$$
$$MSE = \frac{506.19}{48} = 10.546$$

$$F = \frac{MSTr}{MSE} = \frac{58.637}{10.546} = 5.56$$

From Appendix Table VII, $0.01 > \text{P-value} > 0.001$.

Since the P-value is less than α, the null hypothesis is rejected. The sample evidence supports the conclusion that the mean grievance rate is not the same for the four groups.

Since the sample sizes are equal, the ± factor is the same for each comparison.

k = 4, Error df = 48. From Appendix Table VIII, q ≈ 3.79.

$$\pm \text{ factor} = 3.79\sqrt{\frac{10.546}{13}} = 3.414$$

Group	Apathetic	Conservativ e	Erratic	Strategic
Mean	2.96	4.91	5.05	8.74

The mean grievance rate for the strategic group is larger than the mean grievance rate for the other three groups. No other significant differences appear to be present.

15.55   **a** H_o: $\mu_1 = \mu_2 = \mu_3 = \mu_4$

H_a: at least two among μ_1, μ_2, μ_3, μ_4 are not equal.

$\alpha = 0.05$

$$F = \frac{MSTr}{MSE}$$

From the data the following summary statistics were computed:

$n_1 = 6$, $\bar{x}_1 = 4.923$, $s_1^2 = .000107$ $n_2 = 6$, $\bar{x}_2 = 4.923$, $s_2^2 = .000067$

$n_3 = 6$, $\bar{x}_3 = 4.917$, $s_3^2 = .000147$ $n_4 = 6$, $\bar{x}_4 = 4.902$, $s_4^2 = .000057$

T = 6(4.923) + 6(4.923) + 6(4.917) + 6(4.902) = 117.99

$$\bar{\bar{x}} = \frac{117.99}{24} = 4.916$$

SSTr = 6(4.923 − 4.916)² + 6(4.923 − 4.916)² + 6(4.917 − 4.916)² + 6(4.902 − 4.916)²
     = 0.000294 + 0.000294 + 0.000006 + 0.001176 = 0.001770

SSE = 5(0.000107) + 5(0.000067) + 5(0.000147) + 5(0.000057) = 0.001890

$$MSTr = \frac{SSTr}{k-1} = \frac{.001770}{3} = .00059$$

$$MSE = \frac{SSE}{N-k} = \frac{.001890}{24-4} = .0000945$$

330

$$F = \frac{MSTr}{MSE} = \frac{.00059}{.0000945} = 6.24$$

$df_1 = 3$, $df_2 = 20$. From Appendix Table VII, $0.01 > $ P-value > 0.001.

Since the P-value is less than α, the null hypothesis is rejected. The sample data supports the conclusion that there are differences in the true average iron content for the four storage periods.

b $k = 4$, Error df $= 20$. From Appendix Table VIII, $q = 3.96$. Since the sample sizes are equal, the $\pm$ factor is the same for each comparison.

$$\pm \text{ factor} = 3.96\sqrt{\frac{.0000945}{6}} = .0157$$

Storage Period	4	3	2	1
Mean	4.902	4.917	4.923	4.923

The mean for storage period 4 differs from the means for storage periods 2 and 1. No other significant differences are present.

15.57 The following ANOVA table was produced by using the computer packaged MINITAB. Because MINITAB does not compute F ratios, the user must do so from the mean squares which are given. The F ratios are found by the user dividing the appropriate mean squares.

SOURCE	DF	SS	MS	F
Oxygen	3	0.1125	0.0375	2.76
Sugar	1	0.1806	0.1806	13.28
Interaction	3	0.0181	0.0060	0.44
Error	8	0.1087	0.0136	
Total	15	0.4200		

Test for interaction: $\alpha = 0.05$

$df_1 = 3$, $df_2 = 8$. The F ratio to test for interaction is $F_{AB} = 0.44$. From Appendix Table VII, P-value > 0.10. Since this P-value exceeds α, the null hypothesis of no interaction is not rejected. Thus, it is appropriate to test for main effects.

Test for oxygen main effects: $\alpha = 0.05$

$df_1 = 3$ and $df_2 = 8$. The F ratio to test for oxygen main effects is $F_A = 2.76$. From Appendix Table VII, P-value > 0.10. Since the P-value exceeds α, the null hypothesis of no oxygen main effects is not rejected. The data are consistent with the hypothesis that the true average ethanol level does not depend on which oxygen concentration is used.

Test for sugar main effects: $\alpha = 0.05$

$df_1 = 1$ and $df_2 = 8$. The F ratio to test for sugar main effects is $F_B = 13.28$. From Appendix Table VII, $0.01 > $ P-value $ > 0.001$. Since the P-value is less than α, the null hypothesis of no sugar main effects is rejected. The data suggests that the true average ethanol level does differ for the two types of sugar.

15.59

Source of Variation	Degrees of Freedom	Sum of Squares	Mean Square	F
A main effects	1	322.667	322.667	980.75
B main effects	3	35.623	11.874	36.09
Interaction	3	8.557	2.852	8.67
Error	16	5.266	0.329	
Total	23	372.113		

Test for interaction: $\alpha = 0.05$

$df_1 = 3$ and $df_2 = 16$, and from the Anova table $F_{AB} = 8.67$. From Appendix Table VII, $0.01 > $ P-value $ > 0.001$. Since the P-value is less than α, the null hypothesis of no interaction is rejected. The data suggests that there is interaction between mortar type and submersion period. Thus, no tests for main effects will be performed.

15.61 Let μ_1, μ_2, and μ_3 denote the mean lifetime for brands 1, 2 and 3 respectively.
$H_0: \mu_1 = \mu_2 = \mu_3$
$H_a:$ At least two of the three μ_i's are different.

$\alpha = 0.05$

Test statistic: $F = \dfrac{MSTr}{MSE}$

Computations: $\sum x^2 = 45171,\ \overline{\overline{x}} = 46.1429,\ \overline{x}_1 = 44.571,\ \overline{x}_2 = 45.857,\ \overline{x}_3 = 48$

SSTo = $45{,}171 - 21(46.1429)^2 = 45171 - 44712.43 = 458.57$

SSTr = $7(44.571)^2 + 7(45.857)^2 + 7(48)^2 - 44712.43 = 44754.43 - 44712.43 = 42$

SSE = SSTo $-$ SSTr = $458.57 - 42 = 416.57$

Source of Variation	Degrees of Freedom	Sum of Squares	Mean Square	F
Treatments	2	42	21	0.907
Error	18	416.57	23.14	.
Total	20	458.57		

$df_1 = 2$ and $df_2 = 18$. The F ratio is 0.907. From Appendix Table VII, P-value > 0.10. Since the
P-value exceeds α, H_o is not rejected at level of significance 0.05. The data are consistent with the hypothesis that there are no differences in true mean lifetimes of the three brands of batteries.

5.63 The transformed data is:

						mean
Brand 1	3.162	3.742	2.236	3.464	2.828	3.0864
Brand 2	4.123	3.742	2.828	3.000	3.464	3.4314
Brand 3	3.606	4.243	3.873	4.243	3.162	3.8254
Brand 4	3.742	4.690	3.464	4.000	4.123	4.0038

SSTo = $264.001 - 20(3.58675)^2 = 264.001 - 257.2955 = 6.7055$

SSTr = $[5(3.0864)^2 + 5(3.4314)^2 + 5(3.8254)^2 + 5(4.0038)^2] - 257.2955$
  $= 259.8409 - 257.2955 = 2.5454$

SSE = 6.7055 − 2.5454 = 4.1601

Let μ_1, μ_2, μ_3, μ_4 denote the mean of the square root of the number of flaws for brand 1, 2, 3 and 4 of tape, respectively.

H_o: $\mu_1 = \mu_2 = \mu_3 = \mu_4$
H_a: At least two of the four μ_i's are different.

$\alpha = 0.01$

$$F = \frac{MSTr}{MSE}$$

$df_1 = 3$ and $df_2 = 16$.

$$F = \frac{2.5454 / 3}{4.1601 / 16} = 3.26$$

From Appendix Table VII, P-value > 0.01.

Since the P-value exceeds α, H_0 is not rejected. The data are consistent with the hypothesis that there are no differences in true mean square root of the number of flaws for the four brands of tape.